顶级餐厅设计经典 III

北京大国匠造文化有限公司 编

中国林业出版社

·北京·

图书在版编目（ＣＩＰ）数据

顶级餐厅设计经典 ： 全 3 册 ／ 北京大国匠造文化有限公司编 ． —— 北京 ： 中国林业出版社 ，2020.1

ISBN 978-7-5038-9763-4

Ⅰ ． ①顶… Ⅱ ． ①北… Ⅲ ． ①餐馆－室内装饰设计Ⅳ ． ① TU247.3

中国版本图书馆 CIP 数据核字 (2018) 第 224378 号

中国林业出版社·建筑分社

责任编辑：纪　亮　樊　菲　王思源

出　版：中国林业出版社（100009 北京西城区德内大街刘海胡同 7 号）
印　刷：北京利丰雅高长城印刷有限公司
发　行：中国林业出版社
电　话：010-8314 3573
版　次：2020 年 8 月　第 1 版
印　次：2020 年 8 月　第 1 次
开　本：635mm×965mm,1/16
印　张：60
字　数：600 千字
定　价：980.00 元（全 3 册）

Preface · 序

设计指挥家

　　写这篇序言的此刻,刚从美国洛杉矶返回北京。卸下繁忙,坐在工作室一面敲着键盘,一面轻松小酌 Penfolds Cellar Reserve 2007 年的 pinot noir。随着越来越密集的工作行程及频繁的差旅飞行,每年超过 30 个遍及全球的品牌餐饮项目设计,现在我更明白"慢活"对身心平衡的重要,只要工作一有空档,一杯好酒搭配阅读一本好书,代表着一种从容不迫、细腻的生活态度。

　　上月在新加坡发现一间外观抢眼叫"Browhaus"的店,一开始因为发现跟德国"Bauhaus"谐音,走进看竟是一家"修眉"的连锁品牌,专业的技术依据每个人脸型及轮廓,打造最符合个人形象的眉型。就连英国 Wallpaper 杂志也推荐为"国际旅游人士必到的修眉 hotspot"。听说目前上海也有分店,也许下次我能发表变身型男的感受。

　　餐饮空间是人与人聚集交流的场所,所以作为餐饮空间设计者的我,时常会想着什么人来这餐厅,什么是来此的人所需要的,这个过程其实令我感到最有意思,就像是北方人在家包饺子,重点在于包饺子的过程人与人之间的交流,交流的好结果就是水到渠成,交流的不好水饺可能就索然无味了。

　　餐饮空间设计,是一种需要生活经验累积和内在沉淀后的表现。设计工作本身也是有趣的学习过程,通过对生活体验的折射,产生源源不绝的创意灵感并激发出对生命更丰富的热情。像是艺术家或音乐家,基本同样的画笔或是乐器,透过各自精采性格与洋溢才华的表现,总是能撩拨起你我情绪中的那根敏感的神经,而整合了国内外优秀创意的"悦食 3",理所当然的成为整合一切美好设计协奏曲的指挥家。

2013 年 1 月 10 日

悦食

Delightful Restaurant

III

Contents·目录

悦食
Delightful Restaurant III

Contents·目录

Johannes Torpe

Johannes Torpe Studios 设计总监

丹麦著名设计师

The South Beauty Restaurant, Taipei

台北俏江南餐厅

设计单位：Johannes Torpe Studios

项目地点：台北

竣工时间：2012

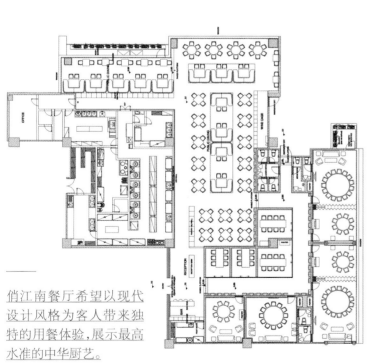

俏江南餐厅希望以现代设计风格为客人带来独特的用餐体验,展示最高水准的中华厨艺。

　　餐厅的家具是由我们和 Johanners Torpe 工作室特别设计的。为了能给客人带来最美妙奢华的用餐环境,我们在设计的各个方面都很用心。

　　厨房对面是一个 LED 显示屏,里面的画面是一个流动的瀑布,透过屏幕你可以看到大厨们忙碌的灶火,整个画面构成了中国传统风水学里重要的水与火两大元素。俏江南的天花板是一幅优美的大型水彩画。它是由许多画家通过数字手段手工绘制而成,总面积442 平方米。置身画下,令人叹为观止。除了这些出类拔萃的设计,我们还为餐厅特别打造了一顶"高帽子",犹如画龙点睛。

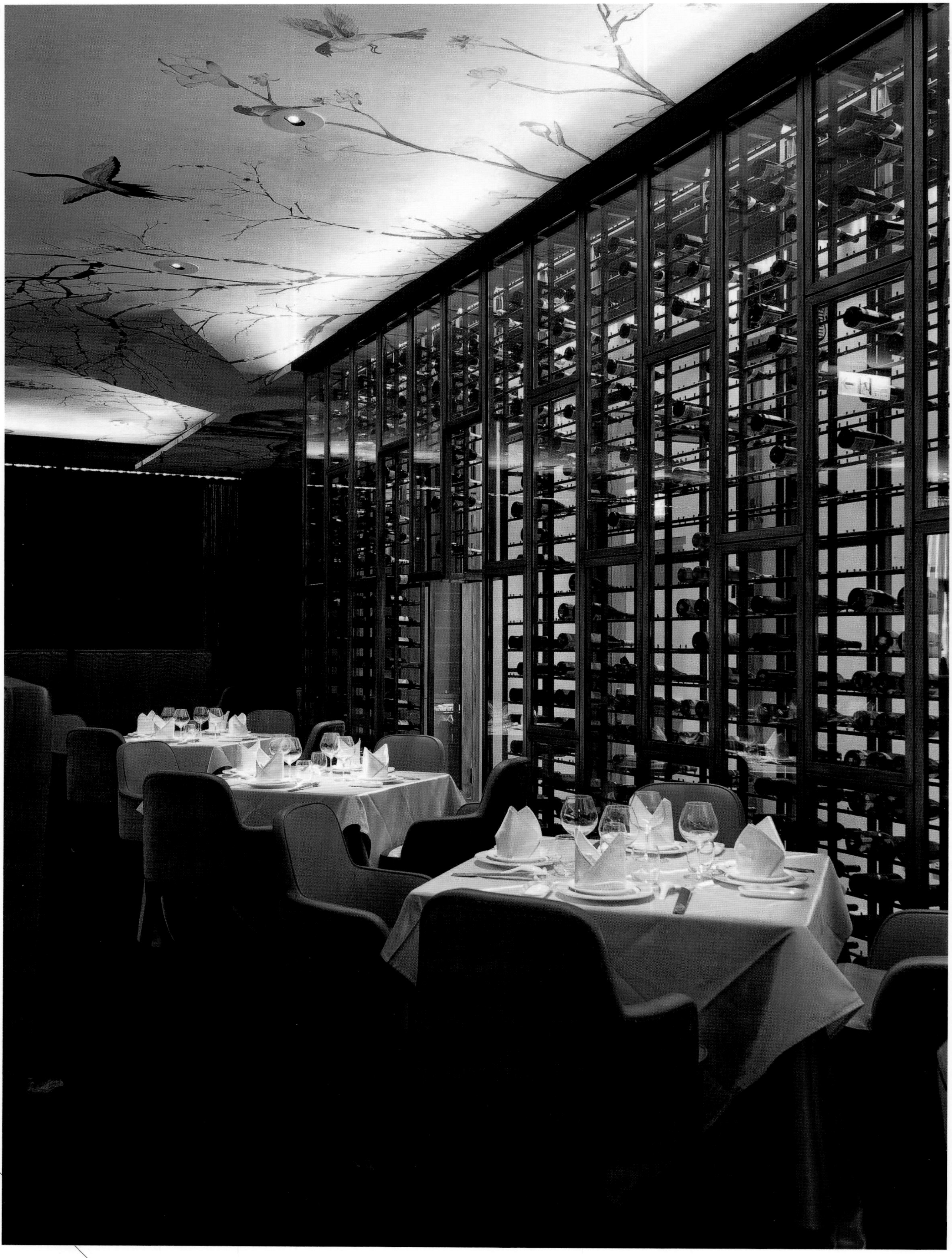

The idea behind the South Beauty restaurant was to create a truly unique dining experience, presenting the highest quality Chinese Cuisine to the world presented in a contemporary design style. We, at Johannes Torpe Studio, designed a special line of furniture to fit in to the restaurant. The vision for South Beauty Taipei was to create a restaurant, which will give the guests the most fantastic and grand dining experience. By designing all aspects of the restaurant it has been possible to do that.

Standing in front of the kitchen is a large LED screen on the entire wall, displaying a video of a continuous waterfall. When looking through this screen into the kitchen you can see the raw flames created by the chefs in action, resulting in the combination of fire and water; two key elements of the traditional Chinese Feng Shui ideology. The ceiling of South Beauty is also very special, bringing hand crafted watercolour paintings from various artists together with digital means to create a 442 square metre artwork that sits elegantly above your head, and contributes to a beautiful and breath taking experience. Together with the ceiling and the rest of the specially designed interior, the cool "high hat" lamps designed exclusively for South Beauty will complete the dining experience.

颜呈勋

教育背景
1999-2003 年 美国 哈佛大学设计学院毕业(建筑系学硕士)
1994-1998 年 美国 杜克大学(经济&艺术史学士学位,优等成绩;优秀生榜学员)

实践经历
2005 年-至今 上海穆哈地设计咨询(MRT design),总监建筑师
2008-2010 年 上海 MRKT 产品,创意总监
2002-2003 年 波士顿 Kennedy and Violich Architecture 事务所,建筑师
2000-2001 年 纽约 Tsao and McKown Architects 事务所,建筑师
1998-1999 年 波士顿 Iconomy.com,设计师

学校职务
2011 年 香港大学(上海分校)讲师
2003 年 上海同济大学 都市规划部 访问学者

Chowhaus at Huashan Rd.

华山路 Chowhaus 餐厅

设计单位:穆哈地设计咨询(上海)有限公司
\MRT DESIGN
设 计 师:Bill Yen\颜呈勋
项目地点:上海
建筑面积:600 m²
竣工时间:2011
摄 影 师:MOSEMAN ELEANOR ELIZABETH

Chowhaus 周边有绿树围绕，门面并不大，木质的外观显得自然低调。

　　餐厅的开放空间被分成 4 个区，右边与中间是适合午餐的座位，左侧沙发、小圆桌适合小酌，一般情况下这就是餐厅开放区的全部，可 Chowhaus 别有洞天。最左往里走，是玻璃房，除却放了不少植物外，中间有个装了壁炉的书架，从各地搜回的老皮箱、旅游手册成为摆设，原色木几周围放着米色、灰色的沙发，自然光线充足，看起来更像个独立的咖啡室。

　　三个从屋顶悬下的巨型玻璃罩其实是音箱，也就是说，在这件玻璃屋内，你可以带自己的音乐来就餐，营造属于自己的小空间而彼此不会互相打扰。相对正式的晚餐区，色调是黑、略深的木色，以及少许金色。内里的两间包房，分别用白色与深木色为装饰，给人以不同色调的冲击感。

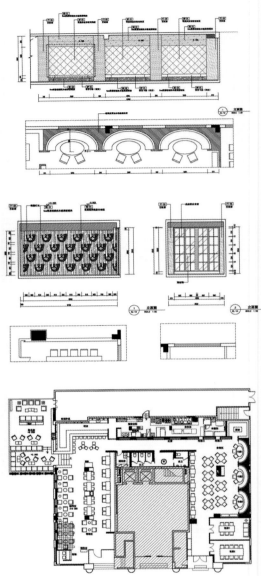

Chowhaus is surrounded by green trees and occupies a small area. Its wooden appearance looks like natural and low-key.

The open space of a restaurant is generally divided into four areas, where the right and middle parts are set with seats suitable for lunch while sofas and small round tables at the left side are arranged for drinking; generally speaking, that is the arrangement of the open space in a restaurant. However, Chowhaus is quite different. Walking inside along the left side, you will see a glass house. There lie many plants and a fireplace-installed bookshelf where old leather trunks and tourist handbooks collected from various places are displayed. The wooden table in primary color is surrounded by sofa in cream and grey colors. With sufficient natural light, the house looks more like an independent coffee room.

Three huge glass covers suspended under the roof are actually sound boxes. So, in this glass house, you can bring music you like to have meals, with your own space made here and without disturbance to each other. The dinner area is relatively formal, where black color and a little dark wood color together with some golden color are designed as the hue. Two compartments inside are decorated respectively with white color and dark wood color, bringing impact sense of different hues.

Being located at the World Financial Center of Beijing's CBD, Cuisine Cuisine Beijing serves top authentic Cantonese cuisine. It's the flagship restaurant of Hong Kong's famous 2-Star Michelin Cantonese cuisine restaurant.

With "being at the center of the center downtown" as the design concept, Cuisine Cuisine Beijing is located at the center of center Beijing city. Covering an area of 3000 square meters, Cuisine Cuisine Beijing features 17 private dining rooms with a total seating capacity of 240 people. Cuisine Cuisine Beijing uses three colors that can specifically represent modern Beijing to demonstrate local specific historical culture. The three colors are gold that symbolizes the culture and history of the Forbidden City, red denoting culture of officials and grey that embodies Beijing's folk life and culture in Hutongs. Being sensitive to culture of China, the designer blends together the three hues which are with great visual impacts and applies respective symbolic meanings of the three pure colors to highlight Cuisine Cuisine Beijing's highlights on cultural inheritance as a Chinese restaurant in modern Beijing. The designer further uses different materials and textures to deepen and highlight respective symbols and features of gold, red and grey. For example, both dignified and graceful gold is used in the massive embroidery

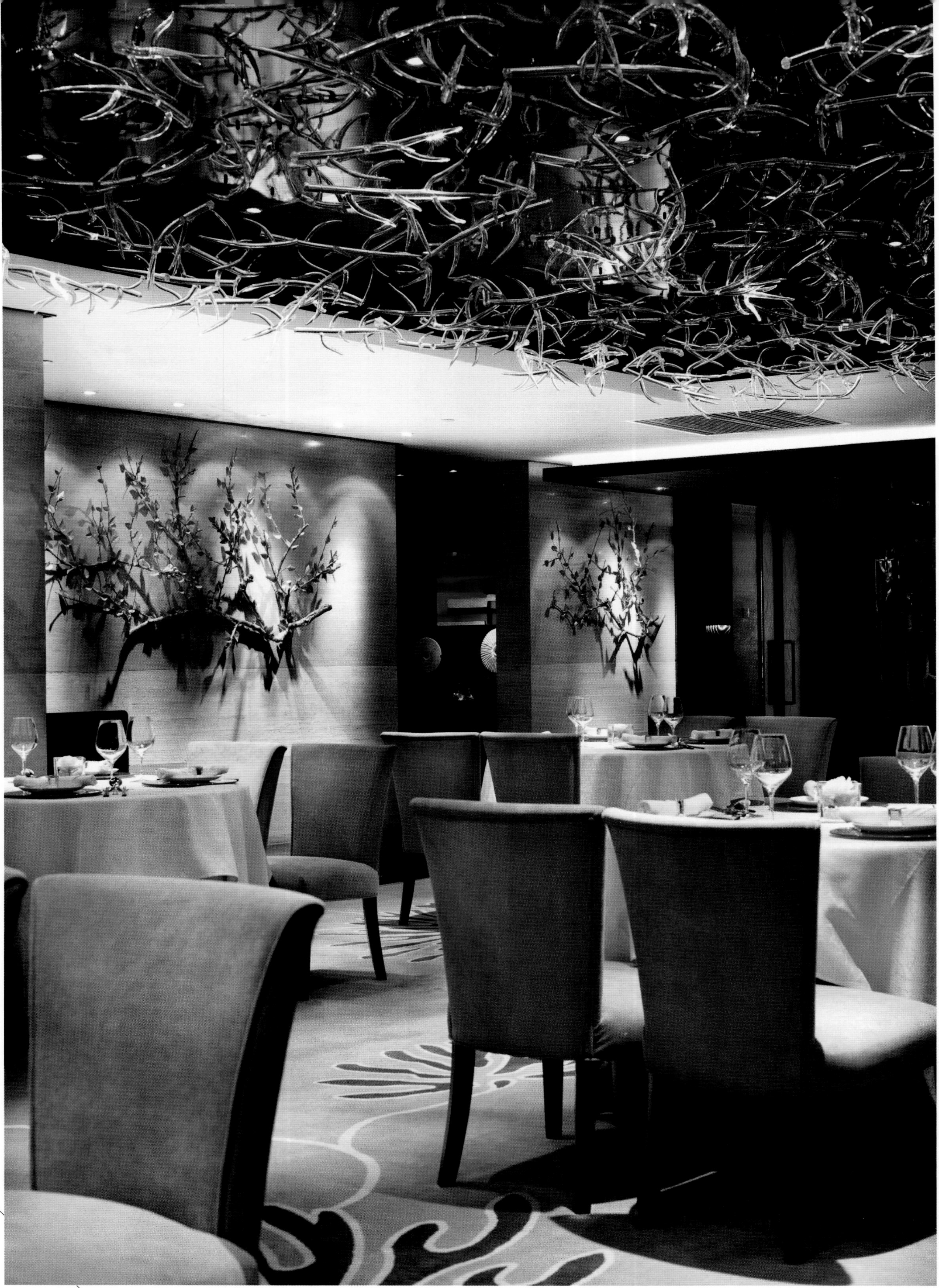

painting on the ceiling of VIP rooms; happy and prosperous red is the color of the whole leather wall of compartments and grey is used to demonstrate defensive walls and traditional quadrangle dwellings, etc. Delicate use of the three main hues brings people strong cultural feeling of Beijing as soon as they step into Cuisine Cuisine Beijing.

In addition, the planning of the restaurant enables guests to enjoy more flexible dining options. For example, guests can choose to dine in a designed opening hall which is like a huge courtyard, or chat and talk about the old days with friends in semi-open space with curtain in a pavilion or select private VIP suite to entertain guests.

Among them the most unique is a "semi-open dining area". The design inspiration of it is from the public activity area in the middle of Beijing's quadrangle courtyard. Golden leaves, red flowers and grey birds on the wall are used for embellishment and decoration and they create a comfortable and homelike dining environment similar to backyard of Beijing's rich family.

Concrete

　　荷兰著名设计公司,Concrete 的整个团队约有35位职业人士:他们是视觉市场商人、室内设计师、平面设计师和跨学科团队中参与项目建设的建筑师。

mazzo amsterdam

Mazzo 餐厅

设计单位:concrete

设计团队:Rob Wagemans, Ulrike Lehner, Marc Brummelhuis, Sofie Ruyten-berg, Femke Zumbrink, Rrik Van Dillen

项目地点:荷兰

建筑面积:400 m²

摄 影 师:Ewout Huibers

Mazzo 是一家意大利风情的餐厅。这是一栋典型的阿姆斯特丹建筑:深而狭窄的空间,天花板的高度不尽相同,地面空间也被分成了好几块。地段也非常适合开餐馆。餐厅的第一部分坐落在玫瑰运河边,这里的天花板高达 5 m,入口朝着运河。这部分最适合做成吧台,客人们可以点一杯速溶咖啡或者在这里喝一杯。灯光采用 GUBI 的 Bestlite 系列,营造了一种令人感到亲切的氛围。靠墙放了一张 9 m 长的低靠背大沙发。墙上和吧台上方的挂灯用了大小不同的两种型号,创造了一种居家客厅的气氛。

餐厅的第二部分矮了许多,但是宽了一倍。几根原石石柱将这块区域分成了 2 个部分。厨房是整个餐厅里最暗的地方,它对面就是用餐区。餐桌的安排简单灵活,也都是餐饮店常用的桌子。餐桌的摆式一般都是 2 人桌或者 8 人桌,每张桌子的位置都能让人一眼看到厨房里忙碌的大厨们。5 组由 MOOOI 设计的 Dear Ingo 灯散发出柔和的灯光,照亮了整个区域。这些现代时尚的枝形吊灯给每一张独立的餐桌营造出愉快而亲切的氛围。墙上挂着 4 幅年龄各异的肖像画,客人在用餐的同时,能体会到意大利家庭式的风情。厨房的隔壁是一间舒适的会议室,里面有一个壁炉,关上门后,客人就能享受到一个舒适隐私的空间。这里适合用来举行商业会议或者家庭聚餐。

餐厅的第三部分向东直达 Bloemstraat 街。这里有一条小巷被改造成了一间密室,白天孩子们可以在家长们的监护下在里面玩耍。到了晚上,玩具会被收到一条黑白相间的窗帘后面。客人们可以靠着一条 11 m 长的靠背沙发上享用美酒,和朋友们聊天。Dear Ingo 的灯光又一次在这里创造了家庭氛围。

作为空间融合多样性以及餐厅自然布局的连接，一个巨大的木质橱柜贯穿了整个餐厅，将所有的空间连接起来，进行统一分配。这个坚实的巨大松木橱柜不仅作为产品陈列柜，还是通往阁楼的楼梯。这里还能作为吧台后方、食物配送点、会议室与餐厅之间的墙、衣帽间、休息室的通道以及儿童玩具的储藏室。

所有这些粗犷真实的设计由五种原料构成：水泥，整齐的砖块，石头，松木以及原钢。前三者取自建筑本身，原钢和松木是新加上去的。

大橱柜的窗架以及餐厅前方的阁楼完全由原钢搭建，这些钢制的梁柱暴露在外。阁楼用未经处理的原钢网围造。这种直接采用原始材料的设计是为了能使客人不被装潢分心，餐厅更专注的是料理本身。

在设计标志时如何融合餐厅的历史和名字掀起了热烈讨论。Mazzo标志是由LED灯在原钢框架上排列组成，让人联想到餐厅的前身是迪斯科舞厅。整个标志如同一顶闪亮王冠悬挂在吧台上方，就算是路过汽车的车灯或者自行车车灯都不能淹没它的灯光。

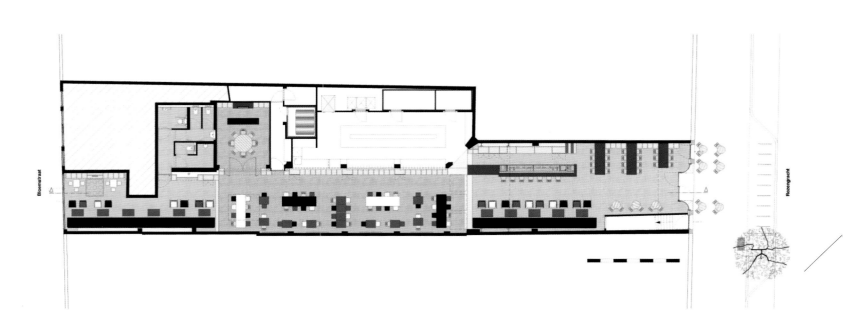

Mazzo is an Italian restaurant. The building is typical in Amsterdam: narrow and very deep spaces fused together with different floor and ceiling levels, automatically providing a natural position for the restaurant.

The first part, situated on the Rozengracht, has a five-meter high ceiling and faces directed outwards. The perfect area for a bar, where guests can order a fast espresso or have a drink at the high bar tables. Lighting from the Bestlite series by GUBI creates the intimacy in this space. The wall lights above the nine-meter chesterfield couch and the suspension lights above the bar and bar tables in xs and XL sizes create a living room ambiance.

The second part is obviously lower but twice as wide. A few original stone columns divide this part into two zones. The darkest part of the entire building is perfect to house the kitchen. The seating area of the restaurant is opposite to the kitchen with a simple but flexible, trattoria-like table arrangement. A variation of tables for two or eight are placed and every table has a great view of the chefs in the kitchen. Five Dear Ingo lights by MOOOI comfortably light up the seating area. Independent of the table arrangement, these modern chandeliers give a pleasant and intimate light. Four portraits of people in different age categories provide an Italian family-feeling of which the guests can be a part during dinner. For a real family dinner or a business meeting guests can use the boardroom. A cosy room with a TV screen and fireplace, next to the kitchen, that can be closed off if privacy is preferred.

The third part is orientated towards Bloemstraat. A narrowing in the building creates the small backroom where kids can play during the day under the watch

of their parents. At night the toys can disappear in the cabinet behind a black-and-white blocked curtain and guests can retreat into the eleven-meter long chesterfield couch to enjoy a good glass of wine and have a nice conversation. Again the Bestlite wall lights by GUBI create a home-like feeling.

The diversity of the fused spaces and the natural restaurant layout need a connecting element: a huge wooden cupboard across the whole restaurant linking all the spaces to each other and organising them at the same time. The cupboard, created out of solid pine wood for storage and display of the products, becomes the stairs to the mezzanine floor, the back bar, point of distribution for the food, the transparent division between the boardroom and restaurant, the wardrobe, the access to the restrooms and the storage for the kids toys.

Five materials determine the ambiance of the raw and honest interior design: power floated concrete, chipped brickwork, stone, pinewood and raw steel. The first three materials are part of the shell of the building; all the new materials are steel and wood.

The window frames in the cupboard and the mezzanine floor in the front of the restaurant are completely made of raw steel. The steel beams and columns are exposed and the extra floor is provided with an untreated expanded metal mesh. The use of honest and simple materials doesn't distract the guests' eye and underlines the fact that the restaurant focuses on the food.

Embracing the past of the building and the name Mazzo by creating a logo that is a controversial part of the venue. The five letters are made of raw steel and filled with classic amusement lights, referring to the disco days of Mazzo. The light object crowns above the bar and will be noticed even in a flash by the cars and cyclists passing by.

MARCELO JOULIA

1958年出生于阿根廷
法国注册建筑设计师和城市规划师
纳索巴黎 & 纳索上海(naco archi-
tectures)设计总监
Architect D.P.L.G. 建筑设计师

Adam Fang 方钦正

毕业于英国曼彻斯特大学建筑系
现任法国纳索建筑事务所设计总
监/建筑师/合伙人

Unico + Colagreco

唯壹餐厅

设计单位:纳索建筑设计事务所
设　计　师:Marcelo Joulia, Adam Fang
项目地点:外滩三号
建筑面积:1100 m²
设计时间:2012.06
摄　影　师:Derryck Menere

"唯壹"提供别具拉丁风情的用餐品酒环境。独特的室内空间来源于建筑设计师的精心设计,并通过独树一帜的时尚设计元素为宾客创造前所未有用餐体验。其中,这些元素包括:设计感强烈,摆放错落有致的古典家具,全上海独一无二的巨型冰雕灯架。此外,异域风情的装饰灯光,摩登复古的地毯,一只叫做COCO的白色鹦鹉作为餐厅的吉祥物当仁不让成为瞩目亮点。

"唯壹"也将与NACO建筑事务所一同建立一个为当代艺术家提供展示他们创新杰作充满活力的平台。从现场表演到由专门的音乐制作人悉心创作精选的音乐,所有的艺术项目都将成为餐厅环境里绝妙的节拍。

"唯壹"Tapas酒廊位于上海外滩三号这个为歌颂"当今的生活艺术"而生的历史建筑二楼。整个空间是由连续的开放式厨房以及令人瞩目的大理石吧台组成。而颜色,造型以及具有独特纹理的装饰材料则配合并呼应着以主要空间为主角的整体氛围。

在"唯壹"内一座绿色的透明玻璃房,里面设置了绿色植物配合180度观赏外滩江景和江畔景观,拥有截然不同的视觉体验。

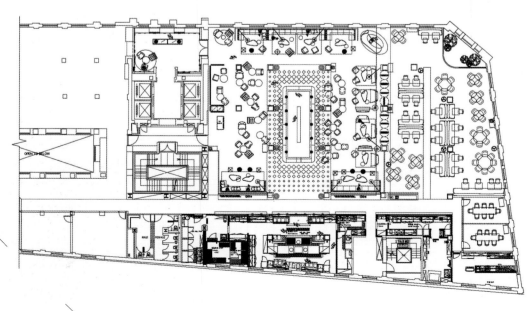

Enlarging Latin culture in a festive space, UNICO interior design is the result of a mature reflexion, illustrated by the most iconic design elements: a patchwork of vintage furniture, a unique ice sculpture lighting suspension, without mentioning decorative lighting, carpets and even a white parrot as mascot called *Coco* ...

UNICO will also become a dynamic platform for contemporary artists to show their creations within Naco Architectures design. From live performances to sophisticated sound selection, the artistic program will give a fantastic tempo to the venue.

UNICO tapas lounge is located at Bund 3 building, a place full of history, celebrating *today contemporary art of living*. UNICO space is organized around the endless open kitchen and impressive central marble bar. Using refined raw materials, the interior design is playing with colors, shapes and textures in an unexpected yet consistent way.

At the waterfront, over the Bund promenade, a green patio, veritable jungle echoing the continent rich nature is creating an extra visual experience.

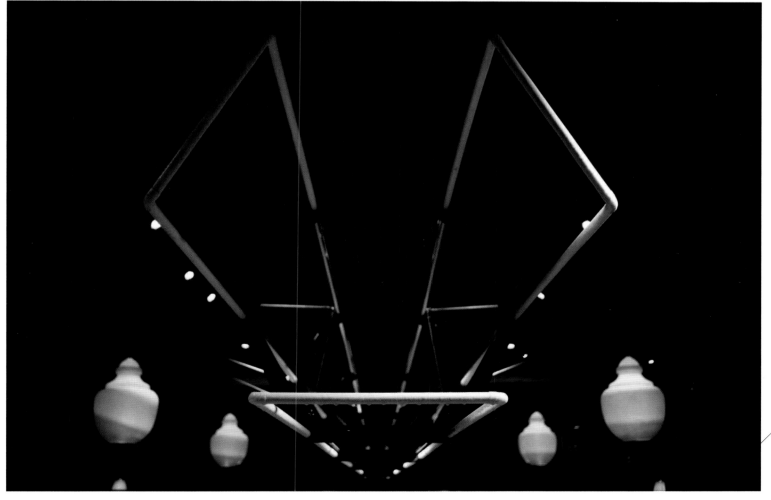

甘泰来

齐物设计事业有限公司总监
交通大学建筑研究所兼任讲师
哥伦比亚大学建筑硕士（荣誉获奖毕业）
康奈尔大学建筑与都市设计研究所研究
东海大学建筑学士

Sabatini-SH

Sabatini-SH 餐厅

设计单位：齐物设计事业有限公司
设 计 师：甘泰来
参与设计：高泉瑜、张芃欣、黄盈华
项目地点：上海
建筑面积：室内 430 m²、室外 94 m²
主要材料：洞石石材、沙比利木、琥珀玻璃、
　　　　　玛莱漆、意大利复古砖
摄 影 师：卢震宇

基地位于商业大楼的地面层,设计师以"盒"状的设计语言,表现空间层层布局的趣味性。

入口处以内凹的方式界定出一座接待门厅,穿越门厅来到了吧台、用餐区、包厢或露台区。特别借由既有的梁柱增建成拱廊造型,让拱柱优雅的轮廓划分出前后段的用餐区域;而层层的门框、拱廊也区分了餐厅的动线转折与座区安排,将各餐桌距离加大,满足交谈隐私以及侍酒与桌边服务所需的空间。考量该区消费客层多为商务人士,设计师除了在入口处规划酒吧作为餐前交谊的场所,更在露台区安排沙发座让宾客能在此轻松交谊。

此外,餐厅特别规划一座VIP包厢,包厢以镶嵌玻璃围砌而成,在水纹玻璃的中介下让内外视景产生朦胧效果,而达到了包厢内部隐私,长型而挑高的空间在玻璃的高透光性而不产生窄迫感。厢房内,可透过一道暗门弹性连结至内部厨房,让贵宾能亲见主厨私房菜的创作过程,包厢也可以再进行场域划分,满足不同人数的餐宴需求。

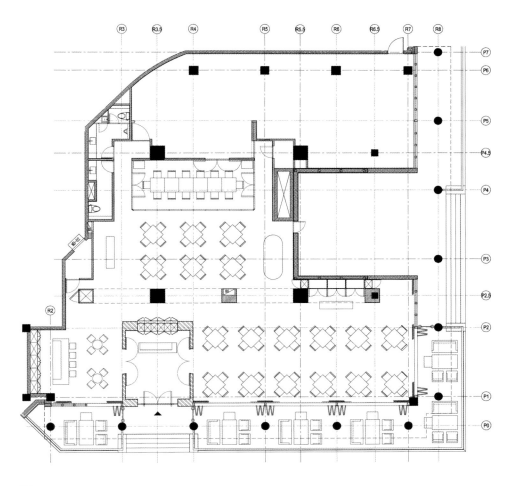

The base is located in the foundation floor of commercial building. The designer uses the designing language of "box" shape to express the interesting of layer upon layer layout.

Within the entrance concave way to define a reception hall, you can across the hall come to the bar, dinner area, box, and balcony area. Special by the existing beam-column built arcade modeling, the counter fort elegant profile divided the dinner area into front and back area. Additionally, Layers of door case and arcade also distinguished between the dining room of dynamic line turning and sitting area arrangement, increased the distance between tables to meet the needs of talking privately, servicing wine and needing space of service beside the table. Considering the customers in that area are businessmen, the designer not only plans the bar at the entrance as the trading area before the dinner, but also arranges the sofa at the balcony area to let the customs communicate relaxed.

In addition, there is a special VIP-case in the restaurant within mosaic glass, in the water cut glass intermediary to internal and external visual produce hazy effect to achieve the compartment internal privacy, long and high space in glass high-translucent do not produce narrow forced feeling. In the wing, through a fly elastic links to internal kitchen, guests can see the chief cook's private kitchens creation process, box can also further divide the field, meet different number of people's meal demands.

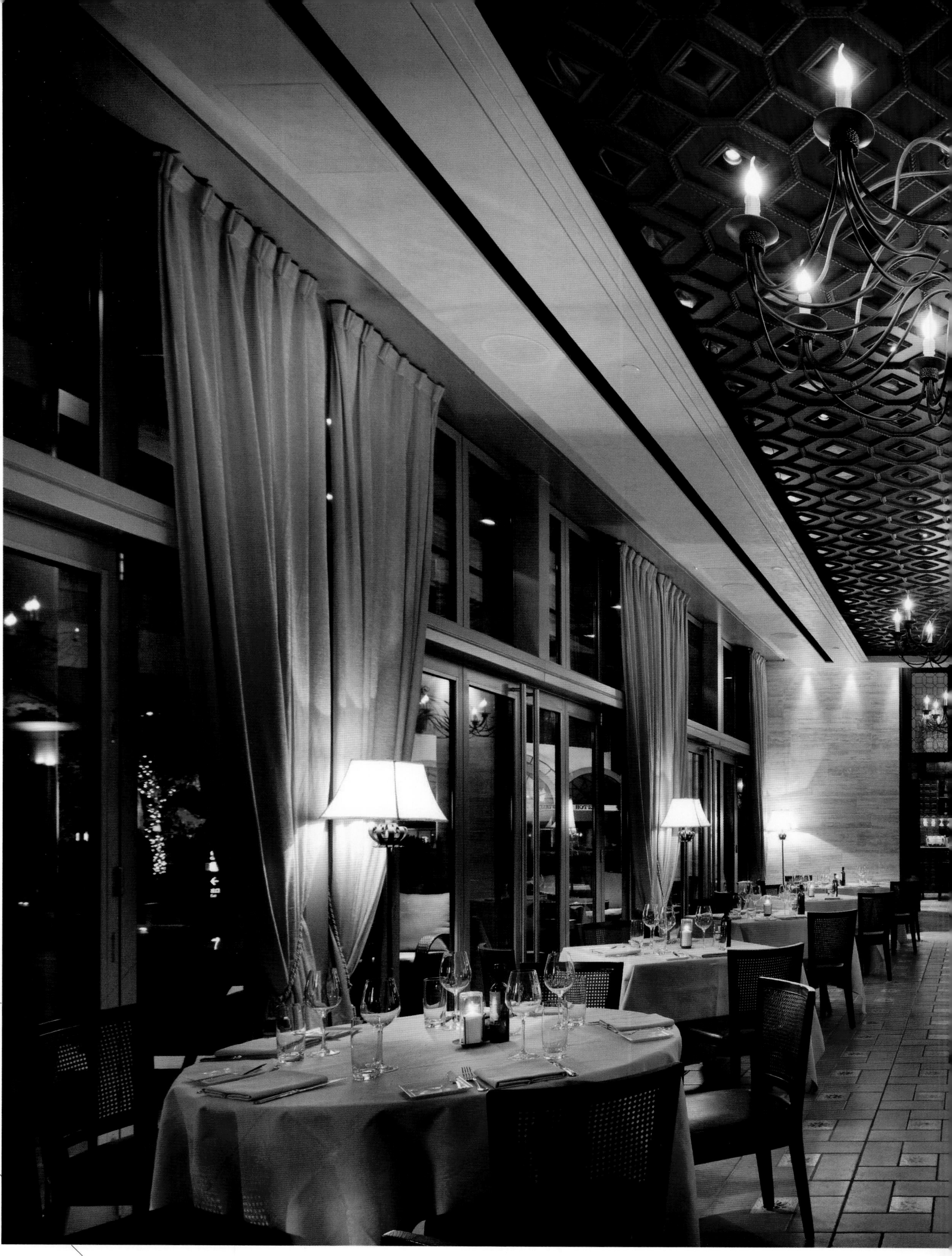

陈德坚 (Kinney Chan)

德坚设计创办人
曾任香港设计中心董事
现担任香港室内设计协会会长

Sesame Japanese Restaurant

Sesame 日式餐厅

设计单位:德坚设计
设 计 师:陈德坚
建筑面积:330 m²
摄 影 师:Trio Photography,
Michael Perini

张永和

建筑师，建筑教育家，非常建筑工作室主持建筑师；

美国注册建筑师；

美国伯克利加利福尼亚大学建筑系硕士；

北京大学建筑学研究中心负责人、教授；

2002年获美国哈佛大学设计研究院丹下健三教授教席；

2005年就任美国麻省理工学院(MIT)建筑系系主任。

Beijing Kingsjoy

北京京兆尹

设计单位：非常建筑工作室

设 计 师：张永和

创意构思：京兆尹

项目地点：北京市东城区五道营胡同2号
　　　　　（雍和宫大殿西侧）

建筑面积：2210 m²

主要材料：砖、瓦、织物

设计时间：2011.08

摄 影 师：舒赫

京兆尹总营业面积2210 m²，可容纳240人同时就餐及举办各类活动。院落环境舒适、风雅，现代科技负离子喷雾迎宾，洗涤一身尘嚣，处于闹市，犹如置身于自然森林中；走进中庭观景院，春揽百花秋邀月，夏沐凉风冬嬉雪；四周金砖铺地，瓦当窗花，人亦融入环境中成为一道风景。宾客在此体验管家式服务、沙龙雅集品茗区、无触碰式卫生间、墙壁上、蚕丝吊灯上处处尽显细节之美，身心也随之轻安。

Total business area of king Join is 2210 m² capable of 240 people repast and holding all kinds of activities at the same time. Amenity and elegant courtyard, greeting guests with anion atomizing by modern science and technology, washing dirt over the whole body, as if place yourself in natural forest though in downtown area. You may accompany by all sorts of flowers in the spring, avoid summer heat in the summer, invite bright moon to enjoy the night in the autumn and play with snow in the winter in the viewing courtyard. Golden bricks cover the ground all around here; eaves tiles take the role of paper-cut for window decoration and people integrate with the environment and form another great landscape. Guests may experience housekeeper service here, while beauty of details is everywhere such as in elegance gathering salon, in tea appreciation area, in no-touch style restroom, on the walls and on the natural silk droplight⋯ Being in the elaborately designed environment, the mood will be comforted accordingly.

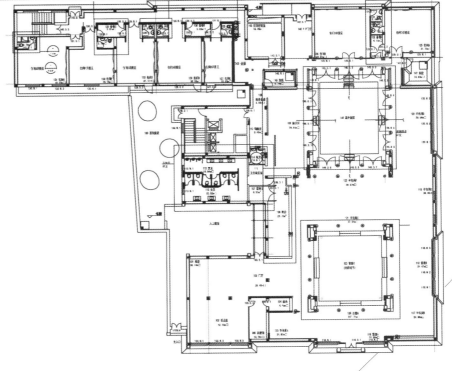

甘泰来

齐物设计事业有限公司总监
交通大学建筑研究所兼任讲师
哥伦比亚大学建筑硕士（荣誉获奖毕业）
康奈尔大学建筑与都市设计研究所研究
东海大学建筑学士

Saburi ATT 4 FUN

纱舞缔 ATT 4 FUN 店

设计单位：齐物设计事业有限公司
设 计 师：甘泰来
参与设计：何智渊、刘煜铃、郑明安
项目地点：台北市信义区 ATT 4 FUN
建筑面积：281.7 m²
主要材料：镜面不锈钢、明镜、茶镜、茶玻、南方松实木染色、柚木木皮、黑色砌石素面石英砖、金属砖
摄 影 师：卢震宇

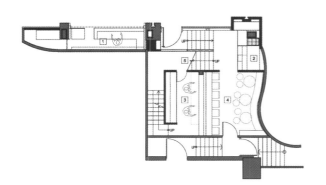

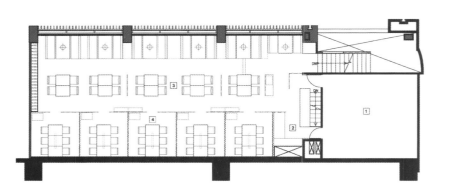

ATT 4 FUN 的纱舞缡餐厅位于台北市精华地段里的信义计画区,挑高两层楼的入口接待区,充满现代设计的空间意象。部分一楼空间除了室内寿司吧台和座位区,还加上提供外卖的黑轮吧台区;楼梯入口利用镜面拉伸橱窗端景的延展效果,天花板矩形不锈钢刻意高低垂吊下来,营造吊灯概念融入空间,木块墙掺杂反射面于其中的凹凸之间;寿司吧则选择实木与石材的搭配。二楼,包厢区则利用茶色玻璃划分出半个开放场域,多为开放的状态下,可透过帘子来区隔用餐空间,依照消费人数的需求调整座位安排。

开放式天花板设计,利用骨料钉制成格子状,于交接处安装矩形镜面不锈钢,产生一个半穿透感的天花板结构,透过不锈钢反射面,模糊掉原本较矮的天花板高度,加上天花板重点的投射照明,其不锈钢镜面巧妙反映着室内光源,不同角度都形成有趣的空间亮点,用餐时让人感到不刺眼。立面墙借由方格实木分别镶嵌在玻璃面上,展现不同层次的丰富面貌,转移原有矮小尺度的焦点,营造模糊视觉界线的空间主题。

Saburi restaurant of ATT 4 Fun, which has an about two - storey high entrance reception area and is full of space image of modern design, is located in the best section of Sinyi planning district in Taipei city. Besides indoor Sushi bar and seating area, a hotchpotch bar which offers takeaway exists in part of the first floor space; the foot of the staircase, which stretch the extension effect of the end view of show window using mirror, the rectangular stainless steel of ceiling deliberately hangs down high-fall to construct droplight concept into space, block wall and

reflector mix in the concave-convex; and Sushi bar chooses to collocate solid wood and stone together.

The open ceiling design, which uses aggregate nail to make into the grid and installs rectangular mirror stainless steel in the connection, to produce a ceiling structure with half penetration sense, plus the projection lighting of the ceiling focus, and the stainless steel mirror which cleverly reflects the interior light source, the different angles all form the interesting space window, these all make people feel not dazzling when they have din-

ner. The gables are inlayed in the glass by the grid solid wood respectively, which exhibit the rich appearance of different levels, and transfer the original small scale focus in order to construct a space theme of fuzzy the line of sight. In the second floor, balcony area is divided into half an open field using the solar glass, under the open condition, the restaurant can separate have dinner space through curtain and they can adjust seat arrangement according to the requirement of consumption number.

利旭恒
出生于中国台湾
英国伦敦艺术大学荣誉学士
古鲁奇公司设计总监

YAKINKUMASTER
Japanese barbecue Restaurant

烧肉达人
日式烧肉店

设计单位:古鲁奇建筑咨询(北京)有限公司
设 计 师:利旭恒
参与设计:赵爽 季雯
项目地点:上海
建筑面积:300 m²
竣工时间:2011.11
摄 影 师:孙翔宇

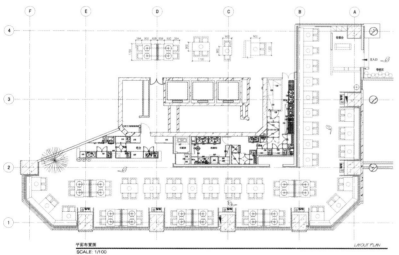

YAKINIKU MASTER 烧肉达人日式烧肉店位于上海天钥桥路上。品牌创立人期望能将日本禅意与中国江南水乡的概念移植到上海,让宾客在舒适优雅的空间里享用美食同时,感受到文化的氛围。设计师利旭恒运用现代的手法演绎日本传统建筑的基本框架结构,大量的木框架朴实地表现建筑结构美学,另外用水墨方式呈现江南水乡中国建筑屋脊的曲线,曲线来自屋瓦,依着梁架迭层的加高,借此强调了中国建筑结构上的不可置信的简单和自然。

YAKINIKU MASTER Japanese barbecue restaurant in Shanghai Tianyaoqiao road, Brand owners aspire to Japanese Zen transplantation with the concept of south China's country style to Shanghai, so that guests enjoy the meal in a comfortable and elegant space at the same time feel the atmosphere of culture. Designer Lee Hsuheng use of modern techniques to interpretation of the basic frame structure of traditional Japanese architecture, a large number of simple wooden framework of the performance aesthetics of building structures, the other showing the curve of the south of Chinese building roof with ink, the curve from the roof tiles in accordance with the beam the rack layer heightening, thereby emphasizing the beauty of this curve in the Chinese architecture construction almost incredible simple and natural.。

陈彬

武汉理工大学艺术与设计学院副教授\硕士生导师

CAAN 中国美术家协会会员

CBDA 中国建筑装饰协会设计委员会委员

CIID 中国建筑学会室内设计分会会员

IAI 亚太建筑师与室内设计师联盟理事

ICIAD 国际室内建筑师与设计师理事会会员

大木设计中国理事会常务理事

后象设计师事务所创始人\设计主持

Design Description of the Inked Orchid Pavilion

水墨兰亭

设计单位:后象设计师事务所

设 计 师:陈彬

项目地点:武汉

建筑面积:649 m²

竣工时间:2011.12

摄 影 师:吴辉

这一餐饮空间是兰亭主题新的尝试，设计的焦点集中在"水墨"一词中，设计师选用拉丝黑钢，白色云石和染色木作，将主色调严格地控制在"黑、白、灰"范围内，决不逾越雷池一步，纯白质地装饰感极强的浅浮雕，将中国传统文化中诗、书、画和古玩以极现代的平面手法表现出来，是空间中的最大看点。最后在完成的空间中选用高纯度的青绿布面的家具，既意外，又合理，流露出设计师对强调入世的中国文人画风的新解。

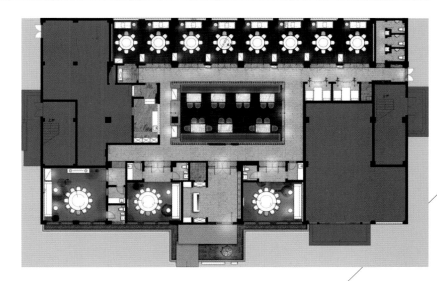

This dining space shows another attempt of the theme of Orchid-pavilion with a design focusing on the word "inked". The designer utilizes the wiredrawing black steel, the white marble and the dyed carpentry, and strictly limits the dominating colors to the range of "black, white and grey" without any overstepping of the limit. The pure white bas-relief with a strong sense of decoration demonstrates the poetry, the calligraphy, the painting and the curiosity in the ancient Chinese culture with modern plane technique, which forms the major feature of the space. Finally, the furniture in dark green of great purity is selected for the completed space, which is quite beyond expectation but meanwhile quite reasonable, revealing the designer's re-explanation to the Chinese scholars' secular painting style.

陆嵘

2002年4月至2006年12月 上海华东建筑设计研究院有限公司室内设计景观设计部主创设计师、主任助理；

2007年1月至2007年12月 上海现代建筑装饰环境设计研究院有限公司第三设计所常务副所长；

2008年1月至今 任上海HKG建筑咨询有限公司中方设计总监和主创设计师

设计的项目多次获得国家及省部级奖励，并获上海十大优秀青年室内设计师及2009年度中国首届中华文化人物称号

蔡鑫

现任上海HKG建筑设计咨询公司设计总监助理、主创设计师

曾就职于上海华东建筑设计研究院有限公司室内及景观设计部、上海建筑装饰环境设计研究院装饰三所

主要设计或参与设计了多个重要大型项目，包括大型办公楼、酒店、会所、宗教的多个领域

Walnew Club
Changguang wetland park, Wuxi

无锡长广溪
湿地公园蜗牛坊

设计单位：HKG GROUP

设 计 师：陆嵘、蔡鑫

参与设计：王文洁、吴振文

项目地点：无锡长广溪湿地公园缘溪道6号

建筑面积：2400 m²

主要材料：松木、红砖、铁锈金属、涂料、清水水泥地面、艺术羊毛毯、刨花木地板

竣工时间：2011.12

摄 影 师：刘其华

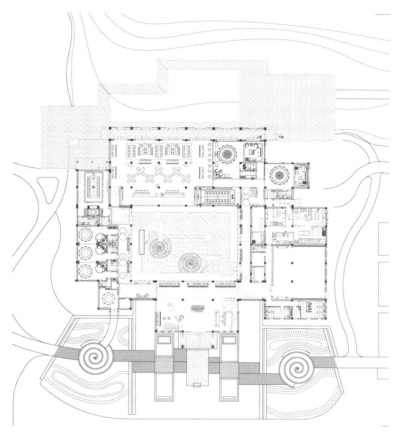

WALNEW CLUB——无锡首家"都市慢生活"的创意餐厅,将设计的创意、美食的享受与湿地的静谧、写意的自然环境融为一体。

结合建筑周围的生态环境,用自然质朴的材料与之相呼应。室内环境基础色调为中性偏冷,通过艺术装置的鲜丽色彩加以点缀,来打破平稳的节奏,从而提升视觉趣味性。家具灯具的设计均简约富有创意,细节之处体现来自大自然中的元素撷取。

室内整体造型线条流畅清晰,虚实有序。无论是墙上块面颜色的铺展,还是顶面的条形格栅造型搭接,始终以简练的几何关系诠释主题的定义:细腻、质朴与自然。

孙黎明

无锡上瑞元筑设计制作有限公司董事设计师

CIID 中国建筑学会室内设计分会理事

CIID 中国建筑学会室内设计分会第三十六（无锡）

专业委员会秘书长

江苏省室内设计学会理事

IFI 国际室内建筑师/设计师联盟会员

ICIAD 国际室内建筑师与设计师理事会理事

美国 IAU 艺术设计硕士

HBO Movie Theme Restaurant

HBO电影主题餐馆

设计单位：上瑞元筑设计制作有限公司

设 计 师：孙黎明

参与设计：陈凤磊、陈贝、陈浩

项目地点：无锡解放西路小尖上2号

主要材料：大理石、电镀不锈钢、绿可木、皮革打印图案、墙纸打印图案、白影木、水曲柳浮雕板、老木头

建筑面积：450 m²

竣工时间：2012.02

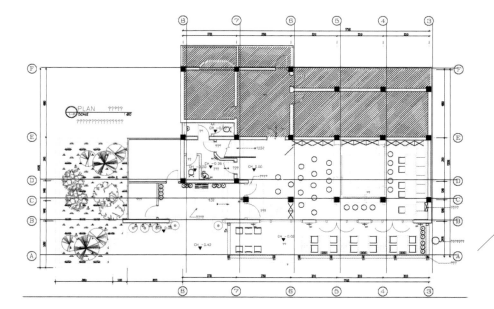

项目业态定位、风格诉求，都源于"HBO"业主陈先生的情结——欧美范儿、热爱经典电影。这是一个能让很多本埠人看得懂的欧美风尚餐厅、一种流溢"奥斯卡"情境的空间味道、一套"做实"的元素组合、一种夹裹怀旧的厚重不失清扬的空间气质，让目标群获得既精致又放松又饶有兴趣的身心体验。空间设计上，充分重视陈设的主表情作用，力求丰富、饱满，而装修则成为背景，为陈设精彩提供恰如其分的舞台——主材的持重含蓄、色彩的肌理自然、结构的大气朴茂。更多的是来自电影视界的细节，海报、胶片、老式电影机、影人肖像、唱片、"那个年代"的自行车等等，有应用要素的呈现，更有平面设计的巧思。

Both project orientation and style aspiration of the project are derived from the love knot of Mr. Chen, proprietor of "HBO". Construct a fashionable European and American style restaurant that is understood by most of local people, create a space with "Oscar" situation, integrate "real" elements and create a place with both nostalgic style and clean & light features to enable target groups acquire delicate, relaxing and interesting mind and body experiences. In the aspect of space design, the designer attaches importance to the main expression functions of furnishings and tries to achieve a feeling of abundantness and plumpness, while decoration is regarded as background to provide suitable platform to furnishings ---- Main materials show dignified and implicit features, while texture of color is natural and the structure is imposing and simple. However, more things are from details of movie industry: posters, films, outdated cin matographe, portraits of filmmakers, records, bicycles of "the age" and the like··· There are the presence of applied elements and more ingenuity of graphic design.

冯嘉云

无锡上瑞元筑设计制作有限公司董事设计师/董事长
中国建筑学会室内设计分会高级室内建筑师
IFI 国际室内建筑师/设计师联盟会员
ICIAD 国际室内建筑师与设计师理事会会员
法国国立科学技术与管理学院项目管理硕士学位

Jiuli River Restaurant

九里河餐厅

设计单位：上瑞元筑设计制作有限公司
设 计 师：冯嘉云
参与设计：陆荣华
项目地点：无锡查桥九里河湿地公园
主要材料：大理石、电镀不锈钢、绿可木、皮革打
　　　　　印图案、墙纸打印图案、白影木、水曲
　　　　　柳浮雕板、老木头
建筑面积：2200 m²
竣工时间：2012.01

步入其间，积极阳光的空间印象顿生。形成这一印象，有对湿地公园大环境的符合，有"低投入、高品质"的甲方预期，更有设计师对"生态"空间的深刻理解与娴熟的手法展现。首先，在平面布局上，在完成餐饮的功能之外，突出了体验型、休闲型、国际感的现代业态气质；其次，材料使用上，环保型材料大量使用，既顺应低碳潮流，又与"湿地公园"保持了精神属性的一致；最后，通过丰富、明快的色彩运用，花卉等植物意向的造型，自然肌理的呈现，勾画出富于生机的就餐环境特质。

Step into the restaurant; customers will be impressed by the ecological, vital and active space which is with a sense of life. This impression accords with the over-all situation of wetland park, shows Party A's prospect of "Low Investment Brings High Quality Outcome" and demonstrates more the designer's profound understanding of "ecological" space and his skilled ways of demonstration. Firstly, in the aspect of layout arrangement, the design fulfills the restaurant's dining function and highlights its modern qualities of experiencing, leisure style and international style. Secondly, in the aspect of material use, the use of plenty of sustainable materials not only complies with the trend of low carbon, but also accords with the spiritual attributes of "Wet Park". And at last, plentiful and vivid colors and plant images such as flowers are used to present natural texture and delineate the features of vital dining environment.

蔡宗志

台湾台南人

台北淡江大学建筑系

法国巴黎建筑专业学院

法惟思设计工作室（Atelier de L'A.V.I.）设计总监

温淼

内蒙古师范大学 环境艺术设计专业

现任 法惟思(北京)建筑设计咨询有限公司设计师

北京宫易锐思国际设计咨询有限公司设计师

Xisi Teppanyaki

西四铁板烧

设计单位：法惟思(北京)建筑设计咨询有限公司、
北京宫易锐思国际设计咨询有限公司

设 计 师：温淼、蔡宗志

项目地点：西四羊肉胡同

建筑面积：1000 m²

主要材料：石材、玻璃、钢构、木材、硅藻土等

竣工时间：2012.03

摄 影 师：孙翔宇

撰 　 写：蔡宗志

在建筑外观上,保留了这一时期的建筑材质,在胡同渐渐消失的现代社会中,能让各时期建筑保留且并存对整个都市的景观有种丰富性,更有其建筑历史发展的意义。处理的手法是不破坏的前提下在外面加上一层玻璃阳光房,使其建筑外观变成室内景观的一部分,还能欣赏户外的四合院。同时为了维持西四胡同区的整体形象,我们保留了门口的形式,做了一些现代手法的细部美化。

在室内风格上,自然粗犷简朴,造景是空间布局的另一个重点,希望让每个顾客在入内的行进间,欣赏不同景观角度,如每日一变的景观水池,大片粗犷的山石,山水纹路的洗石子墙。自然而未过分加工,虚实交叉堆砌出一种静谧的氛围,让顾客在用餐的同时也感受精神上的洗涤。

一楼包间是 vip 包间,力求简单舒适,每个包间都有主题墙,跟烧烤有关。二楼包间以能欣赏窗外的四合院胡同为主,偶尔能看到猫从屋脊上走过,窗外香椿树及松树错落分布在灰色四合院之间,白塔近在眼前。夹层的空间设计尽量不掩饰其原来的木结构,让空间更有个性。

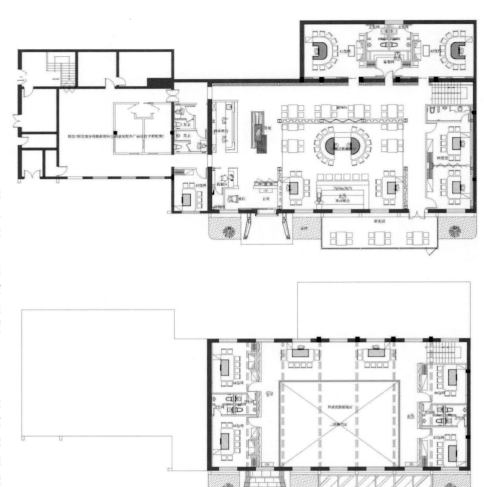

On architectural appearance, we still retain the architectural material quality during this period. In the contemporary society that lanes fade away, there will be a kind of richness in landscape and more significance on the historical development in architecture on the whole city to enable the architecture of all periods to retain and coexist. So, the solution is adding a glass of house outside without destroying, making the appearance on architecture be a part of landscape inside the room , further and decent appreciating the quadrangle dwellings outside houses. To preserve the whole image and in our insisting, the form of gate is maintained. Some contemporary measures on detail beautifying without destroying the globality of quadrangle dwellings will account for the times.

Taking natural and wild style indoors, spatial arrangement is another emphasis on landscaping. In the meantime of spatial experience, hoping that every customer could appreciate the different point of view during the intro-advancing. Like the changing landscape ponds every day, sheets of wild hanging rocks, flushing off the pebble wall with landscape lines, natural but not excessive processing, piling up the quiet atmosphere with true or false crossing ,and thus bring the spiritual washing-up to customers in the meantime of dinner as well as purify the moment of the busy urbanism.

The first floor is VIP rooms striving to be simple and comfortable, and each private room has subject walls related to barbecue. The second floor, focusing on appreciating the quadrangle dwellings and lanes, once in a while you can see a cat pass by on the ridge.Outside the window, Chinese toon trees and pines scattered distribute among the gray quadrangle dwellings, and the white tower is close at hand.

To enable the space to be more individual, prefer not to cover up the original wood structure on the layered space design.

佐佐木力
(SASAKI CHIKARA)

2001 年 3 月 毕业于大阪工业大学工业部

2002 年 3 月 infix（日本总公司）业务研修

2004 年 7 月 上海英菲柯斯设计咨询有限公司

2007 年 4 月 就任上海英菲柯斯设计咨询有限公司
　　　　　　 北京事务所代表

2012 年 2 月 就任上海泷屋装饰设计有限公司副总经理

Starry Night Dining

正村寿司 HK-LokFu 店

设计单位：Shanghai RID Co.,Ltd. /上海泷屋
　　　　　 装饰设计有限公司

设 计 师：佐佐木力

项目地点：香港

建筑面积：130 m²

摄 影 师：Nacasa & Partners Inc.

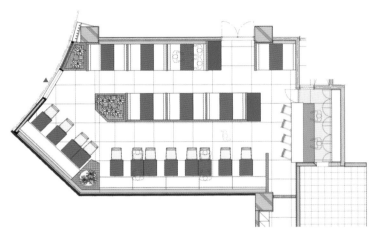

正村是位于香港购物中心内的家常寿司店。

作为日本料理的代表——寿司,在大多数人眼里还是一种门槛比较高的料理,而正村的经营理念就是做让大家在香港就能随意享受这一美食的店铺,营造"和式"的明亮、温馨氛围。

采用3色木材,将体现日本传统风格的格子,与现代风格相融合。格子,不仅能达到装饰效果,还能确保采光和通风,同时在一定程度上阻隔了来自外界的视线,传递传统古建的气息。极富规则性的格子在店内纵横交错、层层叠叠,营造空间整体的动感。而与墙面上的扇形土墙的对比,又突出了空间的进深感。

木材的温和,能将人的感官柔化。我们最大的成就感,来源于能让顾客与亲朋、好友、自己的心爱的人在这里一起度过欢快、愉悦的时光。

Located at shopping center in HK, Zhengcun is a homely sushi store.

As a representative of Japanese food-sushi, within the most people's vision, nevertheless, it has a relative higher threshold, while the management idea of Zhengcun is that it will exactly enable every one to enjoy the good food at will-a good food store, to build the atmosphere both of bright and warm of "He" style.

Using wood which having 3 colors, which will embody traditional Japanese style lattices, mixing together with modern style. Lattices, it not only can achieve decoration effect but also insure both lighting and venting, meanwhile, having obstructed the sight outside to some extent, delivering flavor of traditional ancient architecture. Highly regular lattices cross with multiple layers to build integral innervations of spaces. While contrast to sector cob wall of wall space, it will highlight the feeling both of entering and being deep as well.

Temperate wood will soften sensory organs of humans. It will be the maximum sense of achievement for us to enable customers to spend time both of lively and joyful with relatives, friends, darlings of its own together.

周易

1959年 出生于台中市

1979年 以自学方式学习建筑与室内设计

1989年 创设JOY室内设计工作室

1995年 创设JOY概念建筑工作室

Karuizawa on Gongyi Road

轻井泽公益店

设计单位:周易设计工作室

设 计 师:周易

参与设计:陈威辰

项目地点:台湾台中市

建筑面积:1117.3 m²

主要材料:铁件、铝格栅、文化石、铁刀木、南非花梨木、玻璃

竣工时间:2012.01

摄 影 师:和风摄影　吕国企

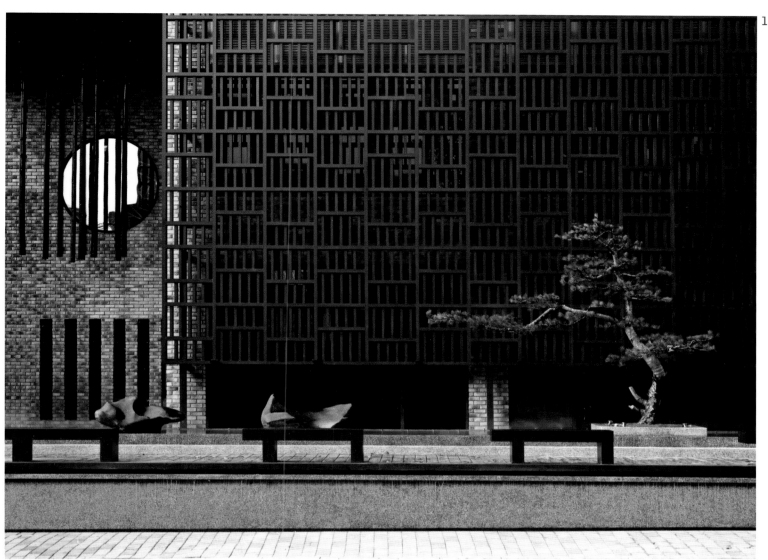

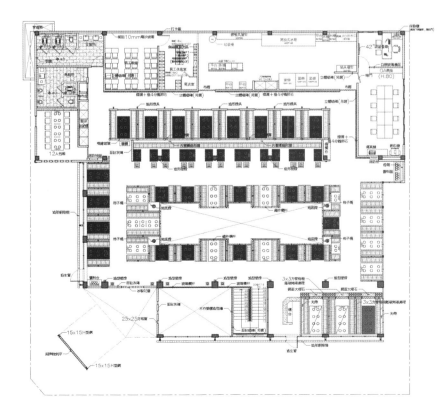

建筑要有意义，不仅在乎工艺内容；更仰赖整体文明、历史、气质的传承。本案延续古朴宏伟的建筑特色，除了静谧禅韵；更多了一份源自悠远中国的人文深度。

基地坐落两路交会的角地，锐角斜切后成为六角形入口与主要店招的展示面，在基地因应地势略行垫高的基座上，超过7 m高的三面黑灰色建筑外观非常有特色，首先是大面铸铁精工打造的倒L型店招＋雨遮，店招正面嵌上书法名家挥毫的巨大白色"轻井泽"铁壳字，夜间在灯光衬托下视觉张力格外鲜明。

沿着架高基座外缘为兼具等待区机能的木栈景观步道，步道与建物之间规划镜面水景，点缀嶙峋的巴东石、烛台灯和蒸腾水雾，并贴心设置别致长凳可供来客小憩。在架高基座外缘与马路的落差处，结合循环水幕与灯光，打造流动而梦幻的光瀑意象，夜幕时分，整座建物仿佛漂浮在光托之上。

静定稳重的建筑主体采用古色古香的斑驳灰砖砌作，两侧立面分别装置大小交错衔接的黑色格栅以及纤长的直列开窗，在宛如古代仓库或碉堡建筑的恒久时间感中，

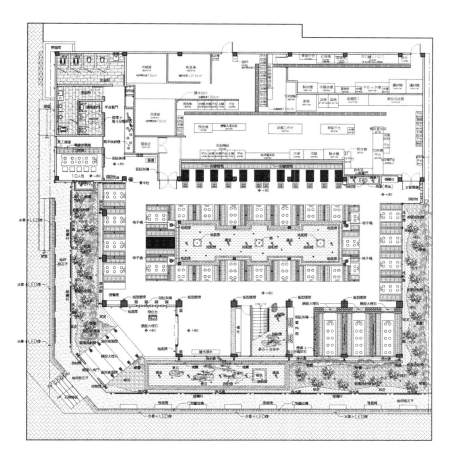

融入中式窗花简化后的格栅线条，低调禅风里，同时摇曳着绿竹的写意以及五叶松的迎宾热情，让这出落在现代的大器建筑，经由设计者绘景叙情的技巧，让画面回荡于飘渺水墨中国与和风浸润之间，反复搓揉出越醇越香的人文风味。

　　入口大门延续外廊的六角门拱，淡雅的苏州庭园意象隐约生成，内部分别为4 m、3 m的二层楼面，分别规划包厢与散座，总计400余的座位，却能避免人声杂踏相互干扰，维持私密又有情调的餐叙氛围。内部空间设定延续大量实木格栅与情境光源，视觉上融入处处可见的古玩书柜墙、大红色喜气洋洋的鸟笼灯饰，形成空间天井并串连上下两个楼层的水景设计，架构三度空间处处有景的绮丽逸界，多处墙面装置运用渐层玻璃与喷绘手法，透过晕染酝酿独特的"光迷幻"效果，柔软而忠实地传达水墨书法与精致图章代言的人文书香，整体层次万千的情境铺陈与灯光布局，透过多种自然材质与高难度工法，凝聚一种跨越时间长廊的精湛演出；当然也让五感在内的享受更有深度。

Meaning of constructions does not lie on-ly on technologies, but more on the inher-iting of comprehensive civilization, history and qualities. This project continues the architectural features of primitive simplici-ty, majesties and also cultural profundity originating from age-old China, besides serenity and Buddhist's charm.

The restaurant which is located at the joint place of 2 roads is featured by hexagonal entrance formed by chamfer-ing an acute angle & displaying surface of main signboard of the restaurant. A building of over 7m with three black grey facades is located at the foundation sup-port raised because of topography and the architectural fa?ade of the building is

with great features. The first features are the reverse L-shape signboard made elaborately by cast iron and the rain aw-ning. The front surface of the signboard is embedded with large,white and ironclad word "Karuizawa" written by calligraphy master. Being foiled by lights at night, the word is with especially bright vision effort.

There is a wooden landscape foot-path with a role of waiting area at the out-er edge of heightened foundation sup-port. Planned water landscape between footpath and the building is adorned with jagged Padang stones, candlestick lamps and rising mist and unique bench-es are elaborately equipped for guests to rest. Zhou Yi, the responsible designer of

the project, once said: light is the soul of constructions. The sentence points out frankly the reality that it is light & shadow that determine the true essence of the circumstance. The most special of the project is the difference of level between the outer edge of heightened foundation support and the road. It integrates circu-lating water screen and light and creates flowing dreamlike images of light water-fall. When darkness gathers, it seems that the whole building is floating above the light.

Peaceful and sedate principal part of the construction is made of old-timey mottled grey bricks and facades at both sides are equipped with black intersect-

ing grilling and long orthostichous windows respectively. It is just like integrating grilling lines simplified from Chinese style paper-cut for window decoration into the lasting sense of time of ancient warehouses or blockhouses or is just like the enjoyableness of green swaying bamboo and white pine's enthusiasm of guest welcoming in silent gentle wind. This enables the modern outstanding construction to express repeatedly the more and more mellow cultural aroma in the immersion of wash painting and gentle breeze by the designer's techniques of scenic painting and passion expressing.

The entrance door designed with hexagonal gate arch as that of outer outline faintly generates quietly elegant image of Suzhou gardens.The interior space is designed into two floors of four-meter-height and three-meter-height respectively. The space is designed into compartments and odd seats which provide over 400 seats in all and the design can not only help avoid mutual interference of voices and noises, but also maintain private and emotional dining and chatting atmosphere. The interior space design uses continuously plenty of solid wood grilling and circumstantial light sources. In vision, the building blends in antique and bookcase walls everywhere & red jubilant decorative birdcage lighting, forms space courtyard and water scenery that connects the upper and lower floors, creates gorgeous leisure three-dimensional space that has views everywhere, applies gradient glass and spraying methods on many wall devices, creates unique "Light Illusion" effect through color infiltration to express gently and faithfully cultural elegance represented by ink painting handwriting and exquisite seals. By various natural materials and highly difficult solutions, the whole circumstance elaborating tens of thousands of layers and light layout creates an exquisite performance covering different periods and of course, brings customers more profound and impressed sense enjoyment.

于丹鸿

重庆朗图室内设计有限公司设计总监

Simplicity restaurant

朴素餐厅

设 计 师：于丹鸿

参与设计：李响

项目地点：重庆市江北区北滨路龙湖星悦荟6-5F

建筑面积：500 m²

摄 影 师：李响

朴素餐厅是一间素食餐厅，提供绿色自然的纯粹素食餐饮，朴素者，天下之大美。这个设计中，试图去营造一个可以让身心片刻宁静，抛开当下的喧扰，可以安静地去寻找每个人本真需求的地方，人和自然达成更和谐的关系。通过设计去讲述一个朴素的生活哲学。

设计意图用尽可能少的设计语言，去表达含蓄内敛的东方文化。期望通过没有具象信息诉求的方式，更多的是暗示和含蓄地表达。收和放，是空间布局的重点。通过起伏蜿蜒的隧道，到豁然开朗的穿顶。从临窗见江的开放明亮，到竹栅围合的私密幽暗。通过收放的布局去引导心理感受的变化。

设计中仅用到两种主要材质，竹和石。江西的细竹片条，制作了围合隔断，隧道，穿顶，地面也是铺贴的同样的竹条。福建的灰色花岗岩石，保留了刚刚开采出来时的劈离石面，无修饰的上墙，通过自然的起伏展现丰富的效果。

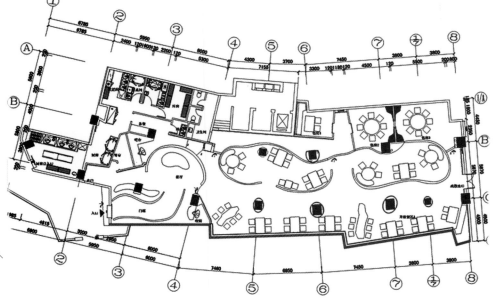

平面布置图

Simplicity restaurant is a vegetarian restaurant, providing pure vegetarian catering both of green and natural, one who is simplicity, which will make world beautiful. The design tries to build a place which would make both the heart and the body be moment's peace, and putting aside immediately pother, which would make one peacefully seek for genuine requirements of oneself, humans would make harmony relations with the nature, telling a life philosophy of simplicity through design.

The purpose of the design would use design language as little as possible to express eastern culture of reserved style. Hope to reservedly express more hints without appeal of concrete information. Both putting away and release is the key to the layout of the space. From winding tunnels which are both ups and downs to domes which make one suddenly enlightened, from both brightness and openness which is nearby a window to both privacy and gloomy which is surrounded by bamboos, to guide the change of mental feelings through the layout both of putting away and release.

There are mainly only 2 materials such as bamboo and stones during design. Fine bamboo chip from Jiangxi had made obstructions which are used to surround, tunnels and domes, and ground is also paved with the same bamboo chip. Ashy granites from Fujian had maintained split stone surfaces, wall run without modifications, which had showed abundant effectiveness through both ups and downs of spontaneous.

潘鸿彬

泛纳设计事务所创始人
香港理工大学设计学院助理教授
香港室内设计协会副会长
IFI 国际室内建筑/设计师联盟执委

Starry Night Dining

星光捌号

设计单位：PANORAMA 香港泛纳设计师事务所
设　计　师：潘鸿彬、谢健生、蔡志娟
项目地点：无锡
建筑面积：700 m²
主要材料：紫红色光纤、枕木水泥壁、生铁、
　　　　　黑白牛图案
竣工时间：2012
摄　影　师：吴潇峰

星光捌号是无锡市中心新开设的一间牛排餐厅,位于历史活化项目"西水东"的百年工厂房之内。设计概念是将富有历史价值的仓库加以改造更新,保留其大部分建筑的原有风貌。加建的夹层结构大大增加了可用空间,使它成为时尚浪漫的现代化餐饮地标。设计策略是将传统的牛排屋餐饮体验提升到一种新的境界。餐厅的两个用餐区内多种类型的座位布局迎合不同顾客的要求,在这时尚而浪漫的意境中享用锯扒之乐。

中庭用餐区保留原有旧建筑的元素,包括10 m高屋顶的斜面天窗和梁柱结构、外墙红砖和室内水泥壁及窗框等,使顾客可以一边进食一边欣赏老建筑的新面貌。2 m高的雄牛屏用刀叉组成,是餐厅的品牌标志,使顾客一见顿生食欲,区内垂直宽敞的空间里,6 m高镜钢铁酒柜和VIP区顶上的水晶吊灯是装饰重点,中央和周围用餐区着真皮梳化,天花垂吊着紫红色管线造成的点点星光,缀成了星光下的晚宴。

夹层用餐区新建的钢构夹层是半开放式。其多功能区的黑白牛图案吊顶和地灯使顾客到此有宾至如归之感。半透明的活动屏风装置提供活动的间隔,以便满足举行不同的活动需要。最后顾客在黑镜饰面的洗手间里结束了其星光之旅,留下了难以忘怀的印象。

StarEight is a new steak house situated at one of the warehouse buildings within a 100-year old factory compound in the city of WuXi, China. A historical re-vitalization design strategy was adopted to preserve most of the original building features with addition of new mezzanine structure to maximise the spatial potentials and turned the old warehouse into a hip F&B landmark. A main scene of "Starry Night Dining" was created to move the traditional Steak House dining experience to a new level of trendiness and romance. Various seating patterns were created in the two dining zones to cater for different customer needs.

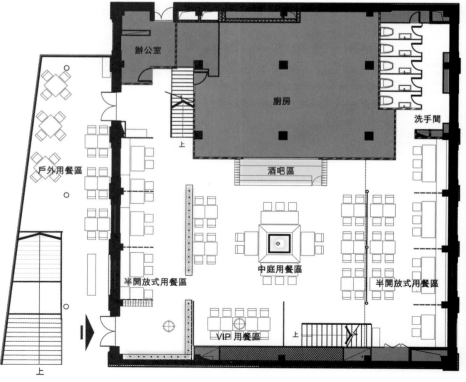

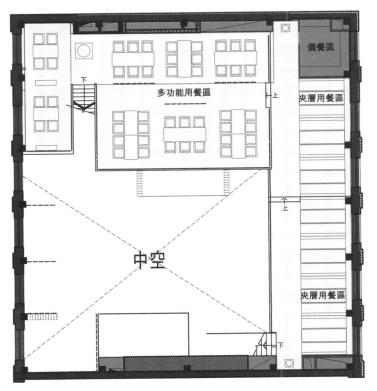

Atrium Dining Zone – Original building features of the 10m high pitched roof skylight, truss structure, exterior brick & interior fare-faced cement walls, window frames were well-maintained and exposed to the eyes of the customers.

2m high bull-shaped partition made by knifes & forks gave clue and defined brand identity to the restaurant. Verticality & spaciousness of the atrium dining zone were emphasized through full-height mirror stainless steel wine rack and crystal chandelier above the VIP table. Ceiling suspended fibre optics in violet colour created starry night effect to the leather-upholstered central & periphery booth seating areas.

Mezzanine Dining Zone - New built steel structure mezzanine floor was introduced to provide a semi-enclosed dining zone. VIP area with black & white bull-patterned ceiling-wall and floor lamps gave cosy & home feelings to the space, sliding semi-transparent curtains provided flexible partitioning for holding different events. The romantic starry night experience was finally completed at the black-mirrored washrooms and left an unforgettable memory to the customers.

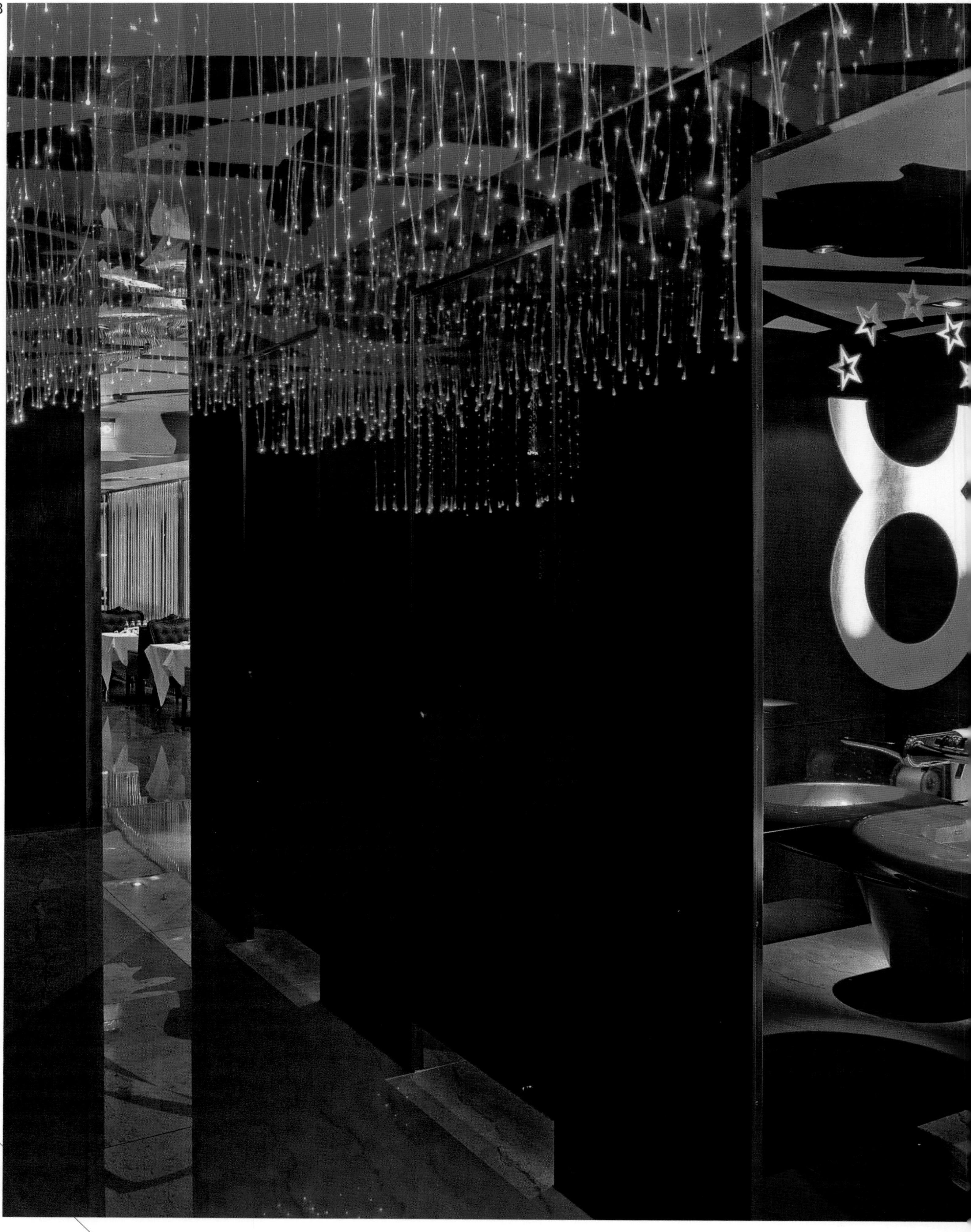

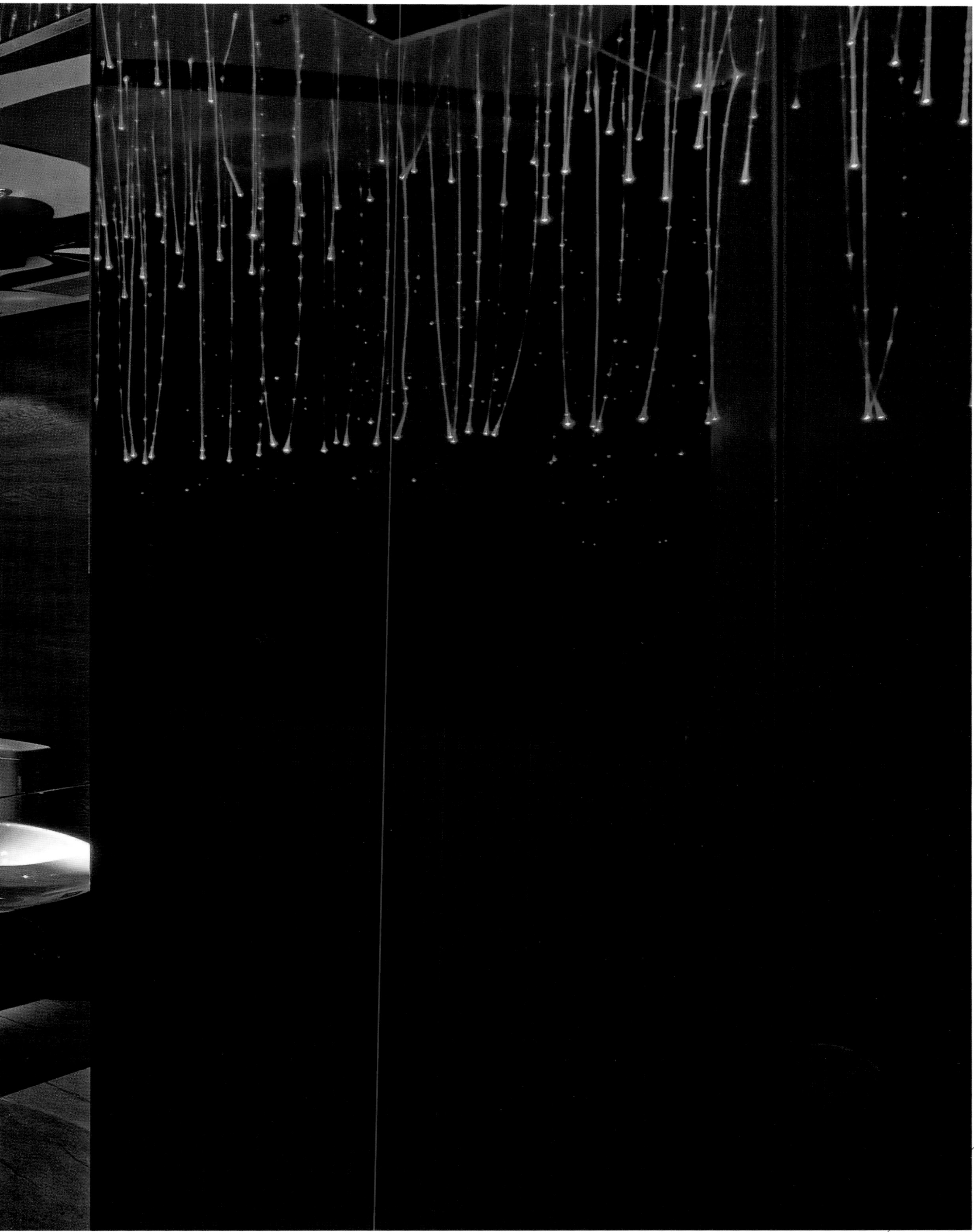

徐晓华

中国建筑学会室内设计分会会员，国家注册室内建筑师。

毕业于苏州工艺美院，苏州徐晓华室内设计有限公司设计总监，擅长于酒店、餐饮、娱乐、休闲等大型娱乐场所的室内设计。

Yangzhou Fulin Hu Garden

扬州富临壶园府邸

设计单位：苏州徐晓华室内设计有限公司

设 计 师：徐晓华

参与设计：苗永光、李爱岭

项目地点：扬州广陵区

建筑面积：1800 m²

主要材料：石材、地板、成品木饰面板、
　　　　　布艺硬包、墙纸

竣工时间：2011.08

摄 影 师：潘宇峰

壶园是一座极具江南特色的私家园林,始建于清代,设计师在其室内设计构思时也结合这一特色,加上现代生活学,给人风雅、舒适的感觉。餐厅地面是深浅纹石材辅以实木拼花地板,深色的圆型柱子加上订制的艺术灯具等复古元素的点缀,整个空间流露出高雅与低调的奢华感。包厢里摆放造型简单、颜色淡雅的家具,复古的陶罐摆设,古典韵味的玉器、贝壳等软装的陈设使一种宁静、高贵、雅致的感觉在整个空间中蔓延。庭院、走廊设计师运用新中式的表现手法,让整个空间又多了一份优雅的文化气息,置身其中,让人流连忘返。

Hu Garden is a most Jiangnan private park. It was built in Qing Dynasty. The designer has merged this character into the space. Coupled with modern life, it gives people a sense of elegant and comfortable. The restaurant ground is made of stone and solid wood parquet floor. The dark round pillars coupled with the customed art lamps and other vintage decorative elements, the entire space seems so elegant and luxury. In the room, there is placed simple elegant furniture, retro pottery and classical jade, shells and so on. They are showing a kind of peace and noble sense, which also extend to the whole space. For the courtyards, corridors, the designer used new Chinese technique to express, letting the space add an elegant cultural atmosphere, making the space full of charming.

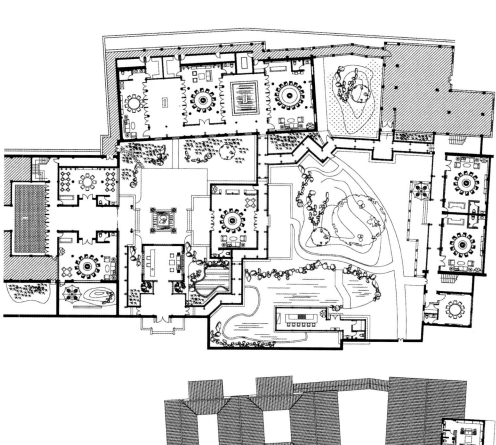

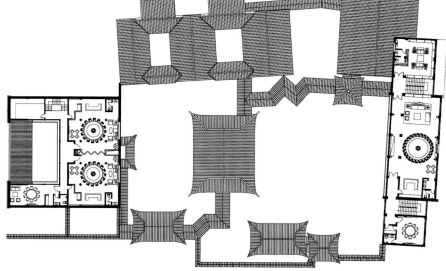

HASSELL

HASSELL

HASSELL 是一家国际化的专业设计公司,拥有 14 间事务所,分布于澳大利亚、中国、东南亚和英国。公司有超过 900 名员工,业务经营已有 70 年历史,涉及全球多类市场。

HASSELL 作为一家跨专业设计公司,拥有多领域专业实力,包括建筑设计、室内设计、景观设计和规划设计,公司注重可持续发展理念与设计实践的结合,兼备卓越的城市设计能力。

作为一家私营合伙公司,HASSELL 的每一家事务所都具备了充分的灵活性和自主性,可同时为本地客户和国际客户提供服务,公司的综合资源和集体专业经验则为各地工作室提供了无可比拟的优势。

2012 年世界建筑设计(BD World Architecture)公布的世界建筑设计公司 100 强中,HASSELL 在澳大利亚设计事务所中排名第一位。HASSELL 已获得超过 650 项设计大奖。

Assaggio Trattoria Italiana

Assaggio Trattoria
意大利餐厅

设计单位:HASSELL
项目地点:中国香港
建筑面积:600 m²
设计时间:2011
摄影师:Graham Uden

Assaggio 餐厅位于香港的艺术枢纽——香港艺术中心 6 楼,面向开阔区域,维多利亚港全景在这里一览无遗。当顾客走进 Assaggio Trattoria 意大利餐厅,即被入口处的比萨与意面吧台所吸引。醒目的艺术作品置于乡村风格石材饰面之上,木纹深色调营造了沉着且愉悦的用餐环境,让人联想起舒适而友好的意大利食品专卖店。意大利文 Assaggio 解作"品尝"或"一口"的意思。设计概念意在打造一个"用餐舞台",充分展现一种意大利杂货店式的舒适惬意氛围。粗糙的烧面砖和深色木地板与色彩鲜明的艺术品让食客暂时摆脱繁忙的生活并进入一个充满艺术氛围的就餐环境。

Being located on the 6th Floor, Hong Kong Arts Centre—art hub of Hong Kong, Assaggio Trattoria Italiana faces open areas and can have full view of Victoria Harbor and gets everything of the harbor at a glance. As soon as customers step into Assaggio Trattoria Italiana, they will be attracted by the pizza & pasta bar counter at the entrance. Striking art works are places on decorated stone surfaces of village style and the dark wood grain creates equanimous and pleasant dining environment and lets people connect comfortable and friendly franchised Italian food stores in mind. The Italian word Assaggio means "taste" or "a bite of". Design conception of the restaurant aims at constructing a "dining platform" and demonstrates fully comfortable and satisfied atmosphere of Italian grocery style. Rough bricks, dark timber floors and colorful works of art help eaters get rid of busy life temporarily and get into a dining environment full of artistic atmosphere. The restaurant enjoys great popularity among customers and eaters since its starting business in March 2011.

Concrete

荷兰著名设计公司，Concrete 的整个团队约有 35 位职业人士：他们是视觉市场商人、室内设计师、平面设计师和跨学科团队中参与项目建设的建筑师。

Hotel Castell D'emporda

castell d'emporda 餐厅

设计单位：Concrete
设 计 师：Rob Wagemans，Ulrike Lehner，Marc
　　　　　Brummelhuis，Sofie Ruytenberg，
　　　　　Femke Zumbrink，Erik Van Dillen
项目地点：西班牙
主要材料：雨伞、白漆刷钢珠、透明窗帘、大理石
　　　　　桌、皮革沙发
图片提供：concrete architectural associates，
　　　　　Ewout Huibers，Robert Aarts

位于西班牙吉罗纳的格泰吉达酒店有一个大露台，在这里可以饱览胜景。露台之上的特色餐厅，客户希望建筑师设计一个封闭的，遮蔽风雨的，但又仅仅像是飘在露台之上，并且与历史建筑和谐的建筑。

　　一般来说，在室外阳台空间享受自然，人们需要一个遮挡风雨的大树，这样流线和视线几乎不受影响。从这个考虑出发，建筑师建立12个雨伞随机放置在阳台上，相互交叠，外层是透明轻质可移动表皮。树立起梧桐树般的形象，轻盈漂浮，与古建筑融合。如果将建筑处理成玻璃屋顶，就像是温室那样，这将与古建筑的融合感大打折扣。阳伞的顶部与侧边做出生锈的效果，与古建筑很搭调，下面是白漆刷钢柱，空间被营造出开放明亮的户外空间感。建筑可以保持开放，但是在寒冷和起风时，也能在1、2分钟之内快速的闭合漂亮透明的窗帘。里面摆放着精心设计的圆形和方形大理石桌，还有白色皮革休息沙发。这个在大遮阳伞下的远行户外酒吧叫玛格丽塔。

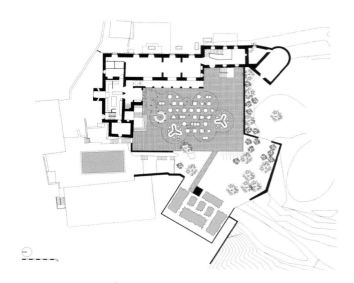

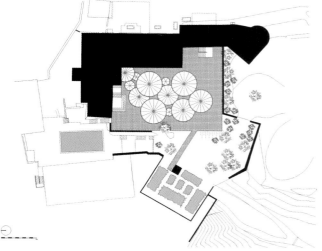

Hotel Castell D'emporda located in Girona, Spain offers a signature restaurant including a large terrace with great views over the surrounding landscape. Concrete designed, at the clients' request, a roof or covering for this terrace with the possibility to create an enclosed space with full wind and rain protection. One of the design conditions was to create a covering that works in harmony with the historical and listed building. Additionally we wanted to maintain the terrace feeling while be seated under the covering.

In principle a terrace is an outdoor space where one can enjoy the weather. If necessary, you need a parasol for sun or rain protection, but there is almost no obstruction between the visitor and the view. The solution was to create abstract parasols. 12 Circles in divers diameters are placed randomly on the terrace. Where the circles touch they melt together, the open spaces between circles are filled in with glass. The circular parasol shapes enhance the feeling of being in an outdoor environment on a terrace. The shape of the covering appears as a separate almost temporary element,

leaving the ancient building untouched. A glass roof or a winter garden would to much become a building, create a feeling being inside a structure and would also appear as an extension of the building, damaging the ancient character. The top and edge of the parasols are made in rust-

ed steel, seeking harmony with the ancient building and the natural environment. The white painted steel columns and ceiling create an open and light outdoor atmosphere under the parasols. Transparent sliding curtains can be hung easily in colder periods but always stay open. When the mistral winds suddenly appear the whole terrace can be closed in a couple of minutes. Round and square marble tables and two white leather lounge couches create different seating facilities. Underneath one parasol a circular outdoor bar is placed. The restaurant now has his own name: Margarit.

蒋建宇

宁波宁海人

2001年组建大相艺术设计公司

2011年组建大相莲花陈设艺术公司

Beijing Financial Street Conti-nental Hotel Xin Rong Ji Restaurant

北京金融街洲际酒店
新荣记餐厅

设计单位:杭州大相艺术设计有限公司

设 计 师:蒋建宇、郑小华、胡金俊、李水、楼婷婷

项目地点:北京市西城区金融街11号

建筑面积:2700 m²

主要材料:京砖、银龙灰大理石、珍珠黑花岗岩、柚木、硅藻泥、铜片

设计时间:2011

摄 影 师:贾方

餐厅位于北京金融街洲际酒店内,地段繁华而高贵。但本餐厅却给人以"自然、空灵、沉静、朴素"之感受,在空间的意境上力求幽玄空灵的精神之美。

在材质的应用上,选择了天然的材料,而且尽可能保留自然材质的天然纹理和质感。比如原木、土砖、竹子、石板、溪石、藤席。利用温润的材料特性营造朴素、内敛的气息,调和人与物、人与空间的和谐。

在空间表达上力求简单、素美,崇尚朴实、自然、亲切,去除过多的无关装饰元素。尽可能减少设计痕迹,使消费者从自然朴素的形态中体验一种幽玄之美,使就餐的空间亦成为人们躲避烦杂世俗的栖息之所。

平面布置图
SCALE 1:200

The restaurant is located in the Beijing Financial Street Intercontinental Hotel, where is noble downtown area. While, the restaurant brings people a natural, ethereal, silent, and pure feeling, as well as, unseen ethereal beauty of the spirit in the space of artistic conception.

In order to retain the natural texture as much as possible, the materials of the restaurant are entirely natural, such as, wood, bricks, bamboo, stones, and rattan mats.

The restaurant makes full use of mildness of materials to create restrained atmosphere and harmony among people, space, and objects.

In the spatial expression, the restaurant tries to show its simplicity, beauty, and gentleness without too many irrelevant decorative elements and to reduce the design traces as much as possible, therefore, making the consumption a journey to experience the unseen beauty from the natural and simple form, as well as, making the dining space a secular shelter for consumers to avoid troubles.

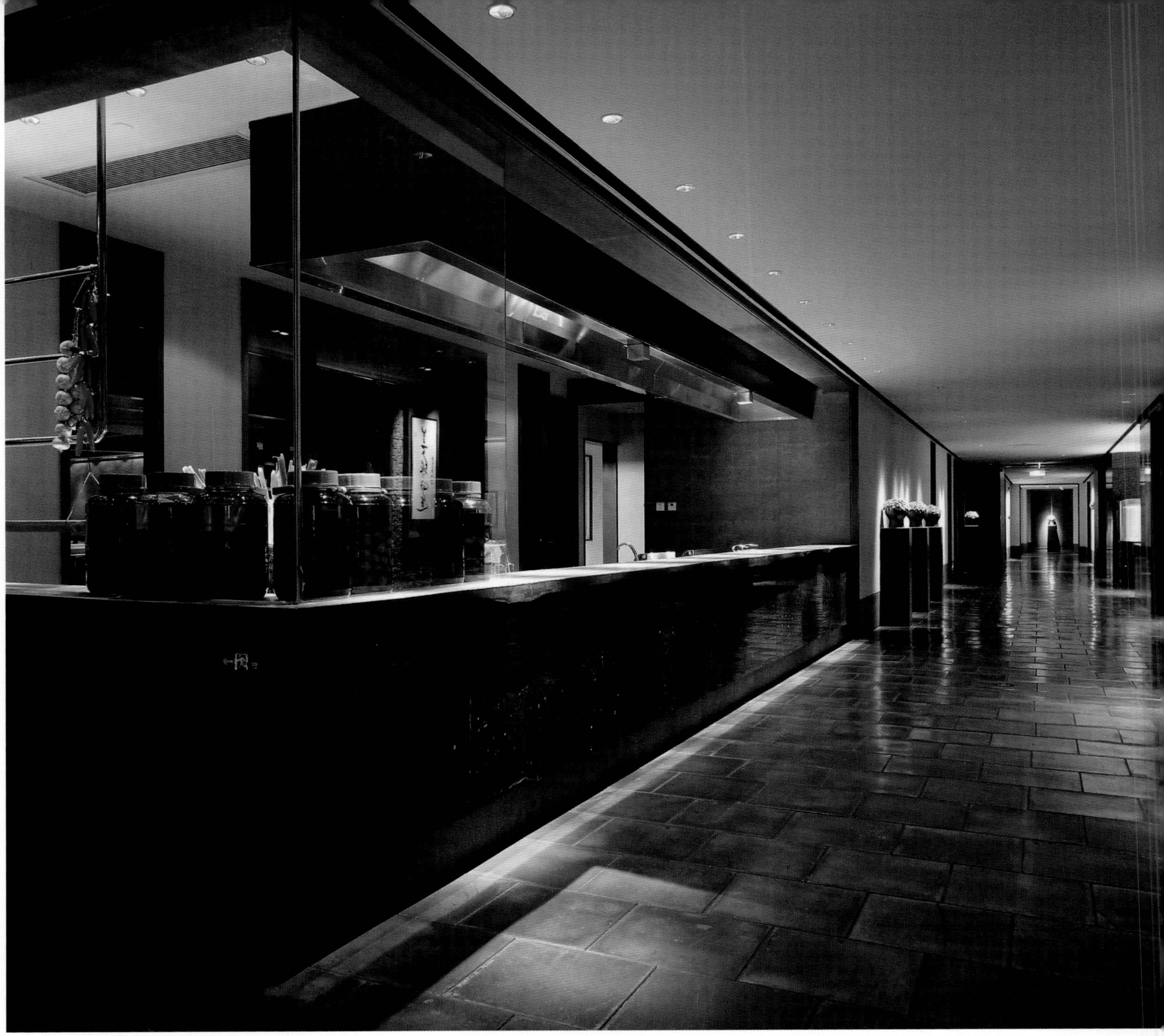

王奕文

和合堂设计品牌创始人\设计总监

蓝犀装饰公司设计总监

Spring feast

春江宴

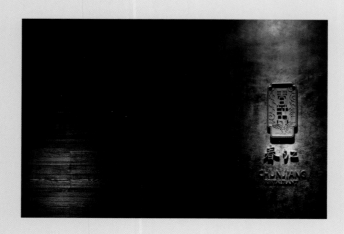

设计单位:和合堂设计咨询

设 计 师:王奕文

参与设计:胡岩峰

项目地点:北京中粮广场

建筑面积:900 m²

主要材料:石材马赛克、地板、壁纸、皮革、烤漆、
　　　　　镜框、绘画作品

摄 影 师:郑尧琛

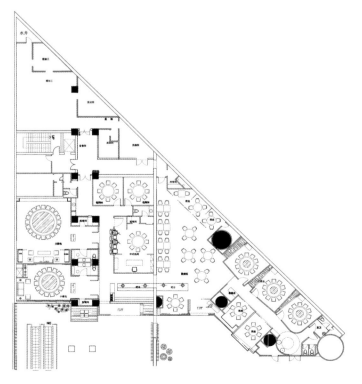

位于北京中粮广场的春江宴,以《春江花月夜》的氛围,营造出优雅,内敛,低调奢华的时尚餐厅。

五个大包间为高端客户量身定做,"江天一色无纤尘,皎皎空中孤月轮",虽无玉雕粉饰,但看似简单平实的灰色,却因为丰富的层次,特别的图案,严谨的配色,震撼的绘画作品,将"低调的奢华"演绎得淋漓尽致。人虽未至诗中景色,然心已向往之。

"鸿雁长飞光不度,鱼龙潜跃水成纹",大厅的鱼灯赋予这内敛的空间些许期盼之意。绿色的座椅,绿色的纱幔,绿色的荷花,暖色的镜框,仿佛镶嵌着一个月光下温暖的画面迎接远行的游子归来。

三个实用性很强的包间为朋友小聚提供了一个安静的所在,当隔断徐徐打开时,那排列整齐的出自艺术家之手的54幅画面,空间尽端的流水坛,轻纱环绕的三个空间印入眼帘。好一副"江流宛转绕芳甸,月照花林皆似霰"的美好意境。

"春、江、花、月、夜"，赋予了这个含蕴，隽永的空间灵动的，深邃的艺术感受。

Located in the Beijing COFCO Plaza, the spring feast makes full use of the atmosphere from the song A Moonlit Night on the Spring River to create an elegant, restrained, and low-key luxury fashion restaurant.

Five large dining rooms are designed for high-end customers, "a sky color river does not have dust, while the moon is bright and silent", although without luxury jade decoration, the five rooms with seemingly plain grey are equipped with rich color level, special designs, precise color matching, and shocking pictures to interpret the low-key luxury to its best. Customers who come here couldn't help yearning although haven't really arrived at the poetry scene.

"Wild goose flies light, diving fish water lines" The fishing lamp of the hall gives the reserved space a bit of hope. The green chair, green veil, green lotus, and warm color frames are just like forming a warm moonlight picture to welcome the arrival of travelling man.

Three quite practical rooms has provided a quiet place for friends reunion. When the cut opens gradually, 54 pictures from artists come into your sight neatly, followed by space water altar and three gauze-wrapping spaces. How nice the atmosphere it is!

"Spring, river, flower, moon, and night" has given the full and meaningful space with smart and deep feeling of art.

利旭恒
出生于中国台湾
英国伦敦艺术大学荣誉学士
古鲁奇公司设计总监

Charme Restaurant A Dream of Dragon at Hongkou, Shanghai

港丽餐厅上海
虹口龙之梦店

设计单位:古鲁奇建筑咨询(北京)有限公司
设 计 师:利旭恒
参与设计:赵爽
项目地点:上海虹口龙之梦
建筑面积:650 m²
竣工时间:2012
摄 影 师:孙翔宇

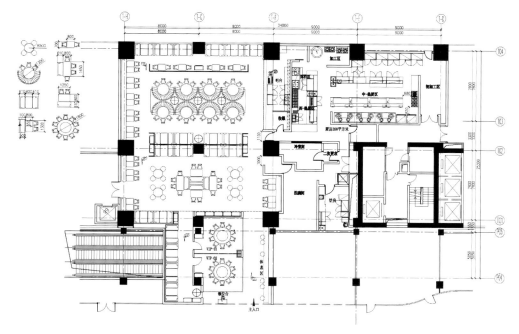

来自香港的港丽餐厅是一家专营港式料理的品牌，在北京与上海都已有大量的粉丝，来自台湾的设计师利旭恒给予了本项目一个有趣的概念："未来世界"。这也是一部40年前的电影。

电影情节进入了空间，机器人布下天罗地网的搜抓人类，人们如同马戏团表演般的在舞台上逃窜，当人类不幸被捕抓之后机器人利用输送带将人类送往另一个世界。

设计师延续未来世界电影概念主题，影射剧情的网状物，输送带，垂直装饰物转化在空间中。从天花一颗一颗的镜面玻璃球，延伸的以多层次灰白基调的墙面造型壁画，对比之后以墨黑皮制座椅作为视觉所及的句点。空间的中央区，吧台造型以未来世界的弧形语汇优雅地呈现科技美学，镜面底板搭配白色人造石桌面构成吧台，简洁的材料语汇，呈现一种未来的时髦美感，吧台的对面是一个大玻璃盒子，内部为条通往地下层的输送带型电扶梯，在餐厅里可以透过玻璃看见双向流动的人群，呈现了未来世界电影情节中的黑色幽默。

设计师为餐厅墙面所制作的装置艺术以"垂直流动"为概念主轴，利旭恒以多层次白灰基调的人造皮革管子，构成一连串的垂直视觉体验，这件雕塑概念的墙面装饰结合了装置艺术的概念，由餐厅入口到用餐区，成为餐厅室内空间的皮肤。

港丽餐厅成功的将电影情节，艺术，设计融入在同一个空间中，成为一体。

Charme Restaurant which is from Hong Kong is a brand that specializes in providing Hong Kong cuisine and it has won plenty of fans in Beijing and Shanghai. Taiwanese Designer Lee Hsuheng gives the project an interesting concept "Future World". The concept is also name of a film produced 40 years ago.

Plot of the film gets into the space and robots spread dragnet to search and grasp human who are like circus performers fleeing in disorder on the stage. When human are arrested miserably, they will be sent to another world by conveyer belts.

The designer continues the concept of Future World and mesh, conveyor belt and vertical ornaments alluding to the plot are used in the space. One and another mirror glass balls on the ceiling and extended multi-level wall of gray tone lead to the full stop of visual sense, the black leather-made seats after comparison. The central area in the space, modeling of bar counter and the arc vocabulary of the future world elegantly demonstrate the beauty of science and technology, while the bar counter consisting of

mirror face baseboard and while artificial stone table and concise materials present fashionable beauty of the future. And the large glass box, which is opposite to the bar counter and whose interior is a conveyor-type escalator and from which bidirectional flows of crowds may be seen, presents the black humor ex-

pressed in the plot of Future World.

Art installation designed for walls of the restaurant by the designer regards "Vertical Flow" as the principal concept axis. Lee Hsuheng uses multi-layer artificial leather of pale grey keynote to create a series of vertical visual experience. This wall decoration with sculpture con-

cept integrates with concept of art installation and forms skin of the interior space of the restaurant' from the entrance of the restaurant to the dining area.

Charme Restaurant successfully integrates plot of the film, art and design into one and into the same space.

200

王砚晨 李向宁

王砚晨（Wang Yanchen）

毕业于中国西安美术学院

意大利米兰理工大学国际室内设计硕士

经典国际设计机构(亚洲)有限公司 首席设计总监

北京至尚经典装饰设计有限公司 首席设计总监

中国建筑学会室内设计分会 会员

李向宁(Li Xiangning)

意大利米兰理工大学国际室内设计硕士

经典国际设计机构(亚洲)有限公司 艺术指导

北京至尚经典装饰设计有限公司 艺术指导

中国建筑学会室内设计分会 会员

Wong's Chafing Dish Restaurant at Yizhuang

王家渡火锅亦庄店

设计单位:经典国际设计机构(亚洲)有限公司

设 计 师:李向宁、王砚晨

项目地点:北京

建筑面积:1700 m²

主要材料:夜里雪、麻砂、做旧实木地板、实竹贴皮、印刷玻璃

竣工时间:2012.07

坐落于北京亦庄的王家渡火锅强调人与自然的和谐。设计师希望走近餐厅的人们能够去感受自然、发现自然，这都归功于现代空间中对传统元素的当代运用。

整个项目的进程既是一次对空间的整合又像是进行着一场再造自然的活动。连续不断延伸的白色石材墙体之上，镌刻的鱼儿带给人们自由与轻盈，白色的天花像微风吹拂的水面，泛起层层波浪，通道两侧大气的通体玻璃表层上，数字化的水墨荷花淋漓酣畅。素白的门板荡起一圈圈的涟漪，而形态各异的鹅卵石，通过不同的材料穿插于空间之中。这诸多的自然意象在室内各个空间之间相互渗透、彼此呼应，通过再造自然空间的手法使环境更加巧妙地融合在一起。

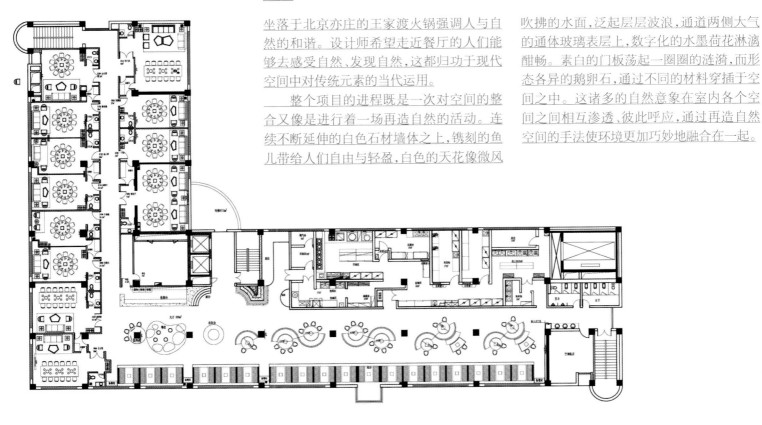

三层火锅平面图

All originates from the respect to the beauty of the nature that requires great unification of sociality and naturality. Wong's Chafing Dish Restaurant which is located at Yizhuang, Beijing emphasizes on the harmony of human beings and the nature. The designer hopes that people who come near the restaurant can feel the nature and discover the nature with their own eyes. And all these are achieved by the contemporary use of traditional elements in modern space.

Process of the whole project is an integration of space as well as an activity of nature reconstruction. Engraved fish on the continuously outspreading white stone walls brings people feeling of freedom and lightness; white flowers like gentle breeze brings waves layer upon layer on the water surface and shows merrily and livelily digitized ink painting lotus on the surface of full body glass at the both sides of passageways. Plainly white door sheet stir up ripples ring after ring, while cobbles of various shapes adorn the space via different materials. These various natural images interpenetrate and act in concert with each other at different places of the restaurant and are integrated ingeniously by reconstruction of natural space.

All of them are selfless presents that the nature brings to us and some sprit consolation that the designer brings to the urbanites that yearn for the nature.

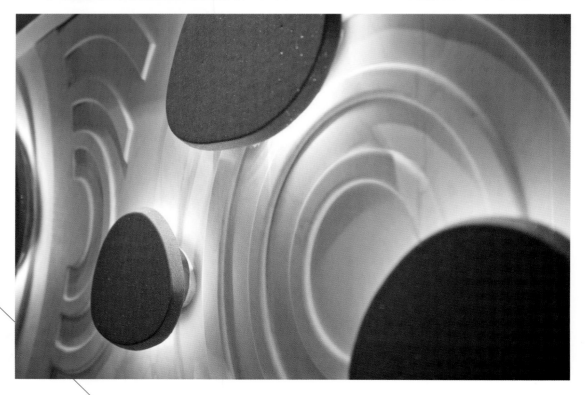

小川训央（OGAWA NORIO）

1963 年 出生于日本大阪
1986 年 毕业于大阪艺术大学
1989 年 加入日本 infix
2003 年 就任上海 infix 总经理
2005 年 就任上海 infix 董事长,总经理
2007 年 成立北京事务所
2008 年 成立香港事务所
2011 年 成立上海泷屋装饰设计有限公司（SHANG－HAI RID Co., Ltd.）

Beijing Longtan Lake Paramount Chamber

龙潭湖九五书院

设计单位 : Shanghai RID Co., Ltd. \上海泷屋装
饰设计有限公司
设 计 师 : 小川训央
参与设计 : 佐佐木力
项目地点 : 北京市崇文区
建筑面积 : 2650 m²
摄 影 师 : 贾方

本次设计，业主十分强调"古典中国的印象"。首先，就中国的历史性建筑物来说，主要的设计有独特的石阶，天花和墙面装饰。采用"在某个部分配合某种装饰"作为设想的基础，然后以此为范本，通过变换展现方式和素材以及其使用方法等进行设计。

　　2650 m² 宽阔的会馆内，几乎由包厢组成，风格多样的设计，能够在不同的包厢享受到不一样的氛围。比如说，天花上吊有艺术品的房间，强调突出天花的高度，通过改变很久以前就存在的中国传统艺术品的活用方式，来塑造室内氛围的变化。除此之外，通过在天花上贴木材等手法，避免"纯粹古典中式"取向的同时，还要特别注意不能设计成现代中式风格。就我们的设计初衷而言，是在"古典"这一设计基调中，同时感受到"当代"特有的时代感。会馆内，也设置了可以摆放古董的展示空间，可以享受到的不仅是店铺的设计，同时还有中国的古典艺术。

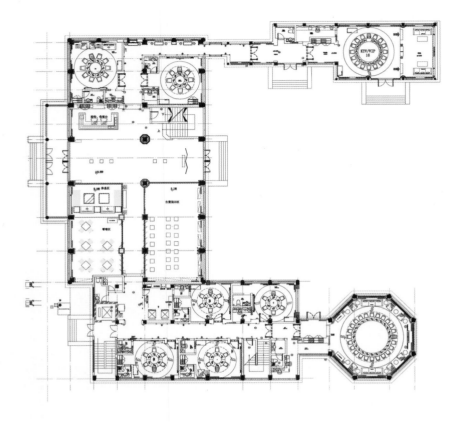

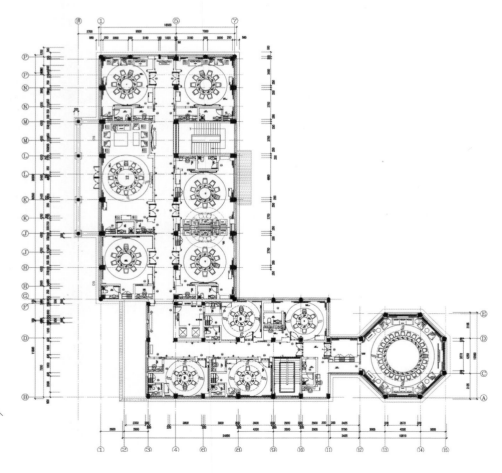

Proprietor of the project emphasizes greatly on "Impression of Classical China", which is a great challenge to us foreigners. However, I thought that it is just because that we are foreigners, we are able to display new and different beauty. Firstly, for Chinese historical constructions, unique stone steps, ceilings and wall decorations are their main design. Therefore, we base on the design concept of "Decorate Certain Parts with Certain Ornaments", regard it as the example and design by conversional displaying and through materials and their application methods…

Almost the whole wide club house of 2650 m² is composed by compartments and design of various styles enables guests experience differently in different compartments. For example, compartments whose ceiling hangs works of art emphasizes on and highlights height of the ceiling. The change of interior atmosphere is achieved by flexible use of traditional Chinese works of art which has been existing for quite a long time. Besides, the action of pasting timber on the ceiling avoids the orientation of "Pure Classical Chinese Style"; however, special attention should be paid to avoiding designing into modern Chinese style. Our original designing intention is to enable guests feel the specific sense of "the present age" in "Classical" designing mood. There is space for displaying antiques in the club house, also. What customers of the club house can enjoy is not only the design of the club house, but also Chinese classical art.

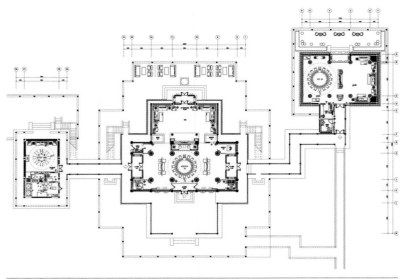

02 一层家具平面布置图@龙吟阁(比例1:100)

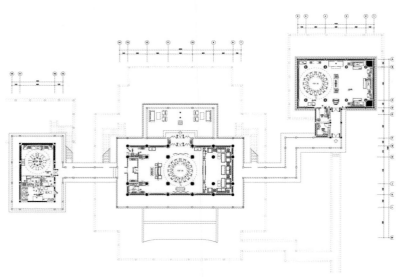

02 二层家具平面布置图@龙吟阁(比例1:100)

利旭恒
出生于中国台湾
英国伦敦艺术大学荣誉学士
古鲁奇公司设计总监

Spice Spirit Restaurant

麻辣诱惑
上海虹口龙之梦店

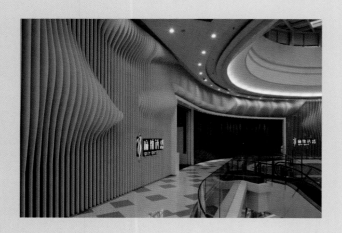

设计单位：古鲁奇建筑咨询(北京)有限公司
设 计 师：利旭恒
参与设计：赵爽、郑雅楠、季雯
项目地点：上海虹口龙之梦
建筑面积：850 m²
竣工时间：2012
摄 影 师：孙翔宇

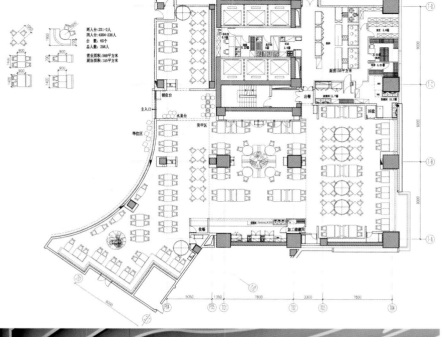

麻辣诱惑位于上海虹口龙之梦，品牌期望将女性体态曲线柔美的移植到空间概念，设计师利旭恒考虑移植原有品牌语汇的同时加入了中国太极的概念，即是在原有阴柔的基础上融入阳刚的多角砖堆砌，利用太极阴阳虚实的关系，寻找一种堆砌与互补的秩序及空间填充的概念。

餐厅用餐空间分割回三个区域，两个元素，一是白色曲线板，板子之间的镜子为了强调太极"阴"与"虚"的女性概念，曲线的构成来自麻辣诱惑品牌LOGO。设计师运用现代的手法演绎品牌形象女性曲线的基本结构，墙与顶面大量的曲线强调了人体美学，横面的曲线来自纵向曲线的迭层，借此强调女性躯体曲线之美，结构上不可置信的简单和自然。

另一元素是紫色多角砖的堆砌，密实的堆砌与大体量表现的是太极"阳"与"实"的男性概念。密实的多角体墙面纵向围合成为中心用餐区域，同时将餐厅分为三个用餐区，这一做法使得扮演男女的主题元素彼此交融，密不可分。

设计团队希望以中国传统太极概念作为基础，使用当代时尚元素具象的表现方式，让宾客在优雅的空间里享受"麻辣"的同时感受到餐饮品牌思想的"诱惑"精髓。

Spice Spirit Restaurant is situated in Dragon Dream Building, Hongkou District, Shanghai. The brand hopes to copy the softness and curved figure of woman's body into the restaurant. While keeping this in mind, Lee Hsuheng also adds Chinese Tai Chi concept into the design, i.e. while reserving the feminine and soft perception, using stacked polygonal-brick wall to present the masculine and firm sense. In this way, he takes the relations of combining masculine and feminine strength in Tai Chi to find the order of piling up and the concept of complementation and spatial filling.

The Restaurant has three dining areas which are decorated in two elements. One of the elements is the white French curves. The mirrors between boards stress the idea of feminine and softness in Tai Chi, and the shape of curves comes from the LOGO of Spice Spirit Restaurant. Lee Hsuheng interprets the brand image of curved beauty of woman's figure with modern techniques: lots of curves used on walls and ceilings highlight the human aesthetics; the laminations created by the vertical curves al-

so produce the curves at the horizontal surface, which makes the unbelievable simplicity and naturalness of woman-figure oriented structure.

The other element is the purple stacked polygonal-brick wall, the dense connection of bricks and the large volume of wall embody the concept of masculine and firmness of Tai Chi. Vertically the wall encloses a central dining volume, and at the same time, divides the Restaurant into three dining areas. This practice finishes the combination of feminine and masculine elements.

The design team wishes to take Chinese Tai Chi as the basic concept and presents it with contemporary design elements and methods, making the customers feel the true spirit of "temptation" while enjoying the delicious "spicy food"

李向宁(Li Xiangning)

意大利米兰理工大学国际室内设计硕士

经典国际设计机构(亚洲)有限公司艺术指导

北京至尚经典装饰设计有限公司艺术指导

中国建筑学会室内设计分会会员

王砚晨(Wang Yanchen)

毕业于中国西安美术学院

意大利米兰理工大学国际室内设计硕士

经典国际设计机构(亚洲)有限公司首席设计总监

北京至尚经典装饰设计有限公司首席设计总监

中国建筑学会室内设计分会会员

Meizhou Dongpo Restaurant,
Yizhuang Store

眉州东坡酒楼亦庄店

设计单位:经典国际设计机构(亚洲)有限公司

设 计 师:李向宁、王砚晨

项目地点:北京亦庄

建筑面积:1700 m²

主要材料:棕云石、银白龙、真丝壁布、印刷丝绸
玻璃、激光切割铁板屏风

竣工日期:2012.07

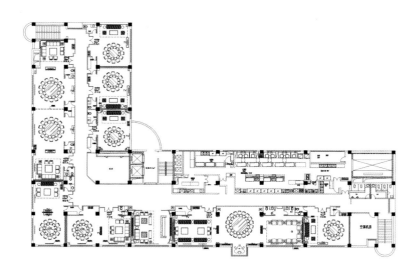

四层平面布置图

眉州东坡亦庄贵宾区项目位于北京亦庄眉州东坡的四层,是基于现已极度饱和的就餐空间新增加的楼层。1700 m²的整层面积只规划了9个包间,最大的包间面积达180 m²,最小的也将近60 m²。品牌创立人期望新的楼层能将博大的东坡文化融入奢侈的空间环境,让宾客在舒适优雅的空间里享用美食同时感受到东坡文化的浓厚氛围。

设计师尝试运用现代的手法,演绎中国传统文化的内在精神本质,展示新材料和传统材料的无限表现才能。利用水墨方式呈现中国传统哲学的处事之道,空间中充满灵动的自然之美。空间中运用大量的传统丝绸面料,与新的玻璃材质相结合,创造出充满自然美又兼顾时代感的材料,优雅地表现传统艺术的温婉雅致。

在四层电梯的入口处,一面仿古铜镜映衬出的画面着实令人叹为观止,出口对面的铁艺雕刻屏风巧妙隔开等候休息区和收银区,传统的中式屏风,由最现代的钢铁材料和激光技术重新演绎,呈现出迥然不同的审美情趣,使空间像充满禅意的中式园林。

空间中大幅水墨山水壁画的载体不再
是传统的宣纸和绢,而是丝绸玻璃,东坡泛
舟赤壁的经典画面也被重新解构,山水和人
物分为两层,在玻璃和丝绸的映衬下,随着
视线的移动,山水和人物在空间中形成新的
视觉印象。这些层迭变化的界面将内部空
间进行了重新定义,顺应了空间跨度的美学
需要,并传递了某种微妙的动感。

Located at the forth floor of Beijing Yi-zhuang Meizou Dongpo restaurant, the project of Meizou Dongpo Yizhuang distinguished guest area is a new added floor based on the extreme saturated repast space now. Only 9 rooms of planning on 1700m2 area of the whole floor ,the area of the maximal room is 180m2 and almost 60m2 of the minimal. The founder of the brand expected that the new restaurant could integrate the vast Dongpo culture into the luxurious spatial environment, offering strong atmosphere of dongpo culture to customers while enjoying the delicacy in the comfortable and elegant room.

Designers had tried to apply the modern skill to perform the inherent spiritual essence in Chinese traditional culture, showing the unlimited performance ability to new and traditional material. Utiliza-tion of ink painting to present the way of the philosophy of Chinese traditional dealing with affairs, fulfilling the flexible natural beauty in the room. Integrated the utilization of much traditional silk plus material in the room into new glass material, which will create the period feel material fulfilling natural beauty and elegant-ly express the gentle and tasteful tradition-al art.

A mirror in the style of the ancient sil-houette against a screen which acclaim as the peak of perfection on the entrance to the elevator of fourth floor. Located at the opposite to the exit, the screen graved in steel art delicately separate the await rest area from cash area.

The traditional Chinese screen is performed again by both the most modern steel material and laser technique, presentation on the widely different aesthetics temperament and interest, which make the room being like the Chinese style gardens fulfilling zen's real meaning.

The carrier of the substantially ink landscape wall painting in the room is no longer traditional rice paper and tough silk but silk and glass. The classic picture of Dongpo yachting around Chibi had been destructed again, dividing landscape and characters into two layers. Against the background both in the silk and glass, landscape and characters had shaped new vision image with the moving of sight line. These interfaces varied in stacks had redefined interior room. Complying with the aesthetics requirements to the spacing of room and delivering some delicate sense of movement, which place yourself not in static room any longer.

朱晓鸣

毕业于浙江树人大学
杭州意内雅建筑装饰设计有限公司 创意总监/执行董事

Mango Thai in Ningbo

宁波美泰
泰国餐厅

设计单位:杭州意内雅建筑装饰设计有限公司
设 计 师:朱晓鸣
参与设计:张天明
项目地点:浙江省宁波市
建筑面积:300 m²
主要材料:定制仿古花片砖、石灰稻草泥、老木
　　　　　板、橡木染色、密度板雕刻、陶土瓦、
　　　　　户外地板、编织木
摄 影 师:林峰

此案位于商场三层,面积并不大,外优势并不明显,注重打造空间的内优势,借以独特的自我特征识别,达到良好的传播是我们设计中考虑的重点。本案在区域划分中反常规地将餐厅入口移置到人流交通的最远端,使来访者在迂回的踱步中对整个用餐氛围有所感知,刻意使客户在餐厅门口廊道中有所积流而不是快速分流,进入等待区后,再通过双通道分流。用餐区根据人群结构不同,割划了对坐区、卡区、散座区、包厢区等。条卡区可酌用餐人数快速拼接来改变其接纳量。而泰式礼品区、水吧区、收银区等整合式"中岛"设计,既减低工作人员的数量,又在视觉交叉点上有极佳全景视线,增加了服务的快速便捷

性。包厢外走廊的过渡空间划分,使其区域有着远离大厅用餐区的视觉错觉感,更加静谧、独立。在设计风格的导入中,考虑其建筑的层高,及结合来访者的年龄层特质,并未一味地将泰式暹罗建筑特质的灿烂辉煌,塔尖翘角等较为异域华丽的元素强加运用,而是在整体较为质朴、平和、随性的"底色"中略施粉黛,恰当地加入了有泰式民族特色的鲜活图案及色块,进行"矛盾"的破坏,如地毯式定制花砖铺设,阵列的多彩门框,绮丽的家具面料运用增加空间的视觉张力;大厅区、包厢区叠水瓦墙的延续运用加以热带植物的烘托,增加其热带南国的风情。借此打破传统泰式风格的贵族仪式感与单调沉闷。空间既有泰国风情又有再创的现代时尚感,创造了充满热带气息的轻松用餐环境。

The restaurant is located on the third floor of the shopping mall and with not large area and comparatively ordinary external advantages. For these reasons, creating the restaurant's internal space advantages and achieving good promotion effect by the restaurant's unique features becomes the important points that should be considered. This case does not follow the conventional regional division, but displaces the entrance of the restaurant to the most remote terminal of stream of people so as to enable customers to have a concrete feeling of the whole dining atmosphere in roundabout entering and intentionally creates the situation of customer gathering at the gate corridor of the restaurant instead of splitting flow quickly. After customers' stepping into the waiting areas, the stream of customers will be split by twin channels. According to the difference of the crowd, the dining area is divided into area for two people, area for group, area for odd customers, compartment area and the like. Area for group may be modified quickly to provide suitable accommodation capacity by jointing. And the integrated design "Middle Island" which integrates Thai-style gift area, restaurant and diner area, cashier area and the like not only help reduces staff amount, but also form excellent full view at the intersection of vision and improves the speed and convenience of the service. The division of transitional space of corridors outside the compartments creates a visual illusion that this area is far away from the dining area and leads to a feeling of serenity and independence. While in the aspect of design style, the designer does

not blindly strengthen the application of the comparatively luxuriant features of Thai-style Siam constructions, such as its magnificence, spire & rake angles and the like, but take floor height of the construction and age situation of customers into consideration. The designer gently decorates the construction in its comparatively plain, peaceful and casual "grounding" and suitably adds fresh and alive patterns and color lumps with Thai national features to carry out "contradiction" violation, for example, use customized carpet-style tiling, colorful array door-frame and gorgeous furniture fabric to improve the visual tension of the space. The continuous use of small cascade and tile walls at the hall area and compartments area are decorated with tropical plants to increase its tropical austral amorous feelings and break Thai-style's sense of noble ceremony and monotonous depression. The space not only has Thai amorous feelings, but also recreated sense of modern fashion and light dining environment with tropical atmosphere.

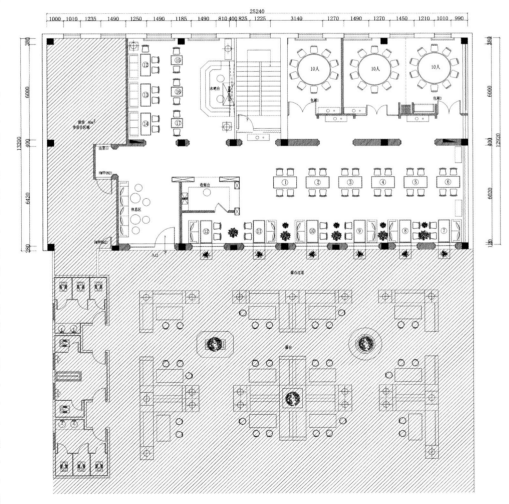

一层平面图
Scale 1:80

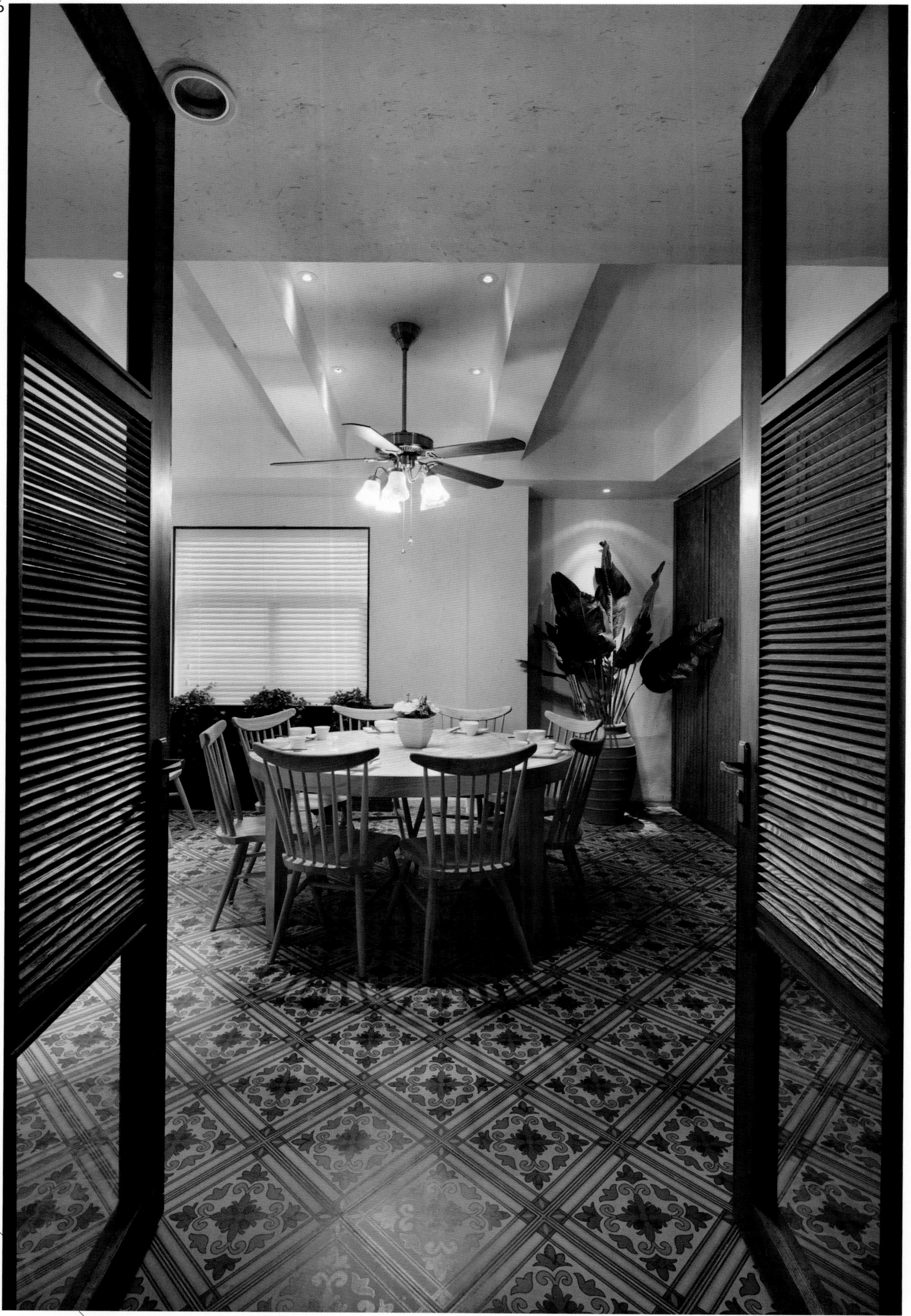

熊华阳
深圳华空间机构总经理\设计总监
中国建筑学会室内设计学会会员
高级室内建筑师

Club House of Honor,
Tongren, Guizhou

贵州铜仁
上座会馆

设计单位：深圳华空间机构
设 计 师：熊华阳
项目地点：贵州
建筑面积：7000 m²
主要材料：大理石，绿可木，布艺沙发，装饰画

坐落于贵州省铜仁市的上座会馆,是接待当地高端人群及高级客商的重要场所。在这样一座有山有水的自然静寂之城,最适合于它的建筑莫过于传承我们几千年历史文化的中式风格建筑了。上座会馆由两座三层高的中式建筑砌合而成,由建筑外观到室内设计,及软装配饰等,均由华空间一体化完成。会所内含有健身会所、娱乐酒吧、中式餐饮等项目。

设计师在设计项目时需扬长避短,充分利用项目的优势,上座会馆依托于当地的山水之景及少数民族特色,所以从外观设计到室内设计,使用中式的设计框架结合现代风格的家具、饰品,院子中央的小池塘、少数民族特色的壁画、现代风格的沙发、中式古典的木椅、竹叶图案的地毯……使会所由内及外散发出传统的高雅的新中式设计风格。

空间的设计不在于将墙面及吊顶做复杂的处理,而在于给它恰到好处的点缀;古典中式的沙发是否会有些呆板? 我们结合现代简约的沙发一同陈设,流畅的线条即显现出来;普通的过道如何才能拥有设计感? 我们使用镜面并带有图案的玻璃做墙面;正统的中式包房拥有古典的中式家具就足够了吗? 我们设计了与众不同的玄关增加了包房的设计感;您是否忽视了楼梯的设计? 但我们注重项目的每一处细节,选用几何形状的时尚楼梯扶手,并在楼道陈列着艺术品展示……

无论是室内设计,还是产品设计,成功的设计在于相辅相成,由点及面的互相呼应。如此案中多处应用的方圆结合,会馆外观,接待大厅,池塘边上,包房内的玄关,都是方圆之间的艺术组合。

Club House of Honor which is located in Tongren City, Guizhou Province is an important place for receiving local high-end groups and high class guests and businessmen. To a natural and quiet city with both mountains and water, the most suitable construction for it is Chinese style construction that inherits our historical cultures of thousands of years. The Club House of Honor consists of two three-floor Chinese style constructions. All things of the club house, from the fa?ade of the constructions to their interior design, soft decoration & ornaments, etc., are completed comprehensively by Hwayon. The club house contains items such as body building club house, entertainment bar, Chinese style repast⋯

While designing for projects, designers should adopt the project's good points and avoid their shortcomings and take full advantage of the projects. Relying on local landscape and minorities' national features, the Club House of Honor integrates Chinese style design framework with modern style furniture, decorations, small pond in center yard, murals with national features of minorities, modern style sofa, wooden classical Chinese style chairs, carpets with bamboo leaf patterns from the design of fa?ade to interior design⋯ Therefore, the club house is able to express fully traditional and elegant new Chinese style design style from the interior to the exterior.

Space design does not lie on complex treatment to wall surface or suspended ceiling, but lie on ornament to the point. Will classical Chinese style sofa seem a bit rigid? Display it along with modern simple sofa will highlight flowing lines. How to make ordinary passages with graphic sense? Let us make mirror planes and glass with pictures the wall surfaces to settle the problem. Will it be enough to equip orthodox Chinese style compartments with classical Chinese style furniture only? The creating of extraordinary hallway will enhance the graphic sense of the compartments. And have you ignored the design of staircase? We pay attention to every detail of the project and for staircase we choose fashionable geometrical shape stair railing and display works of art at the passageway⋯

No matter it is for interior design or product design, the successful ones should be those whose parts can well supplement each other and whose points and surfaces take concerted actions, such as this case which integrates square and round in many places: façade of the club house, the reception hall, pond side, hall ways in compartments... All of them are art combination of square and round.

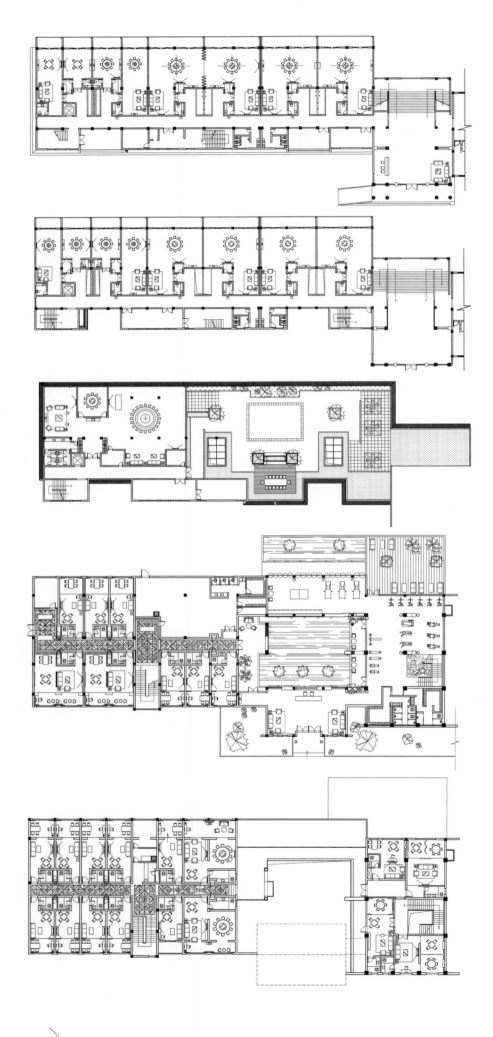

陈彬

武汉理工大学艺术与设计学院副教授\硕士生导师

CAAN 中国美术家协会会员

CBDA 中国建筑装饰协会设计委员会委员

CIID 中国建筑学会室内设计分会会员

IAI 亚太建筑师与室内设计师联盟理事

ICIAD 国际室内建筑师与设计师理事会会员

大木设计中国理事会常务理事

后象设计师事务所创始人\设计主持

Design Description of the Favorite Pavilion

所好轩(沌口店)

设计单位:后象设计师事务所

设 计 师:陈彬

项目地点:武汉

建筑面积:360 m²

竣工时间:2011.12

摄 影 师:吴辉

所好轩（沌口店）是主题店中店的再次演义，设计师依旧将色彩选定为空间主角，继续延续红与蓝的主旋律，在灯光的营造下，上演了一出空间梦幻剧。

特别定制的漆面仿纸折皱钢板天花，从网速缓慢而被延伸的图片中获得灵感的红色透隔，点缀着传统道家星宿图的丝绒皮制墙板，模仿自然晨雾和雨丝的特效隐光玻璃被巧妙地组合在一起，使空间充满戏剧性，并传达着设计师以这个时代特定的方式向传统文化的敬意。

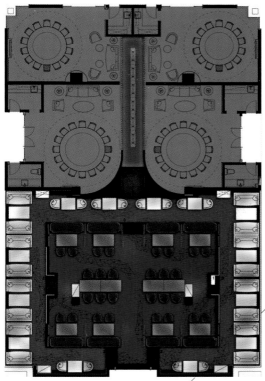

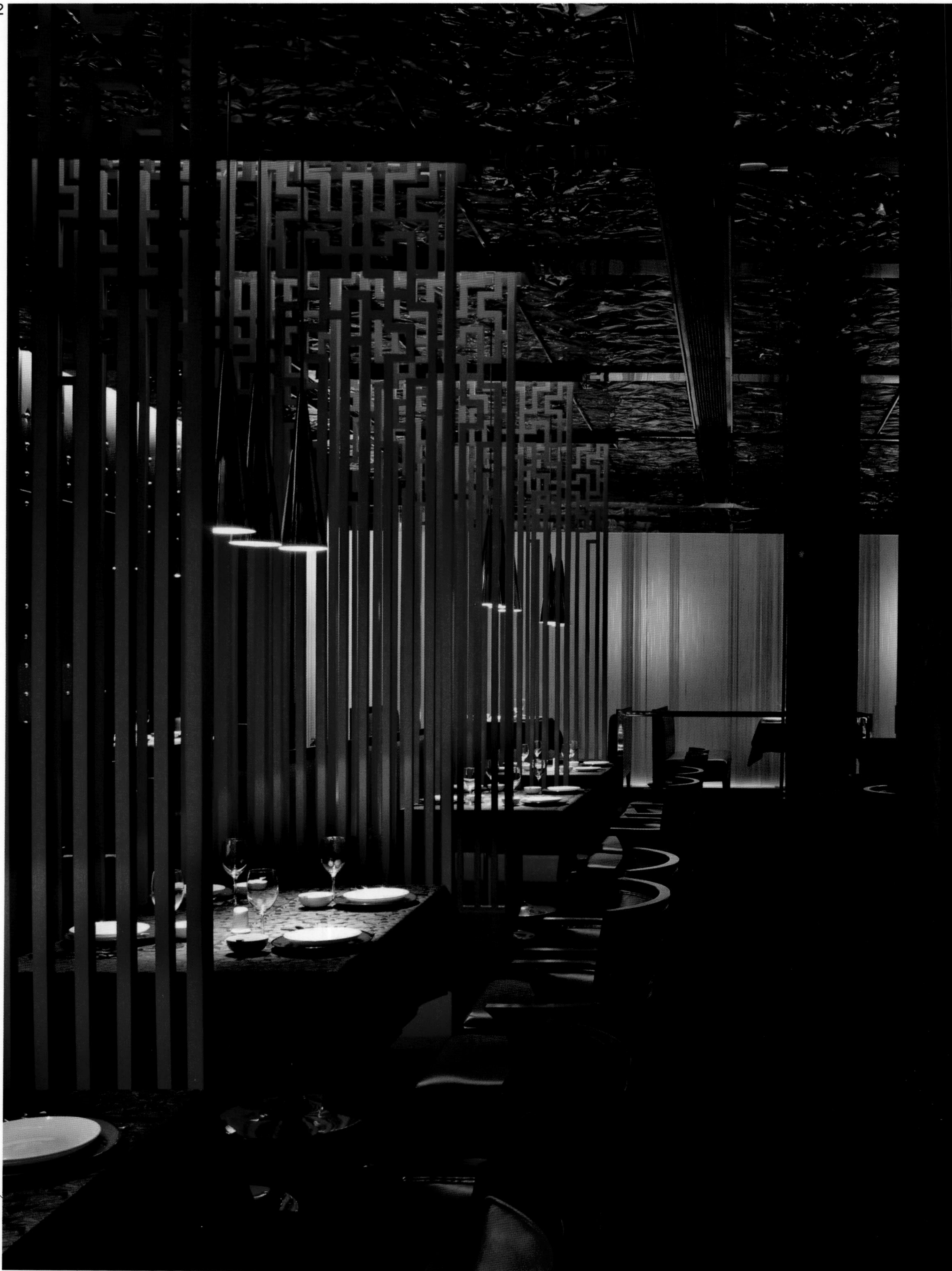

Favorite Pavilion (Zhuankou) is another demonstration of the themed restaurant in a main restaurant. The designer still chooses colors as the major role of the space with an extension of the theme of red and blue color. The lighting effect creates an atmosphere of a spatial dream play.

The artistic combination of customized ceiling of synthetic lacquered paper and wrinkled steel plate, the red transparent heat insulation inspired by the extended pictures due to slow network speed, the velvet leather wall board dotted with the constellation drawings of the traditional Taoism and the frosted glass imitating the effect of natural morning mist and drizzle creates the dramatic feature of the space and expresses the designer's respect to the traditional culture in a specific way of the contemporary era.

王奕文

和合堂设计品牌创始人\设计总监
蓝犀装饰公司设计总监

Bianyi workshop

便宜坊

设计单位：和合堂设计咨询
设 计 师：王奕文
参与设计：赵光宇、胡岩峰、吴林国
项目地点：北京华润五彩城
建筑面积：1400 m²
主要材料：建筑灰砖、木雕、壁纸、金漆、镜框、绘画作品
摄 影 师：毛立广

灵感来源于雍容华贵、富丽端庄的牡丹，本案是和合堂设计在"便宜坊"连锁店中的又一创新，寄予其牡丹花卉图案雕刻、牡丹颜色寓意等手法，运用布局、色调、灯光、材质等来演绎富贵吉祥，繁荣兴旺的氛围。

入口 8 m 挑空位置为富丽大气的金色门面提供了不可多得的空间条件，弧形收银台的二龙戏珠实木雕刻，背景 8 m 的木隔断，红色写意绘画作品，无处不渗透着高贵典雅端庄秀丽的视觉冲击力。

与入口相邻的区域为半开敞包间区，设计师赋予其牡丹花瓣般的平面造型演绎，完成空间的半封闭状态，灰色轻纱飘渺间，红色灯笼造型吊灯若隐若现，仿佛亭台楼阁间的婆娑倩影，令用餐空间变得更为趣味性和戏剧性。

顺着动线的指引豁然开朗，大的散座区映入眼帘，天花延续了花瓣圆形造型，此区域可容纳 200 人同时就餐，为婚宴、大型聚会提供场所。

二层是浓墨重彩的包间区，狭长的弧形走廊，一侧为歌颂便宜坊历史的金色词句雕刻，一侧为以牡丹的色彩为线索而设计的五个包间，黄色、红色、绿色、白色、墨色的半开敞空间将略显深邃的通道变得情趣盎然。

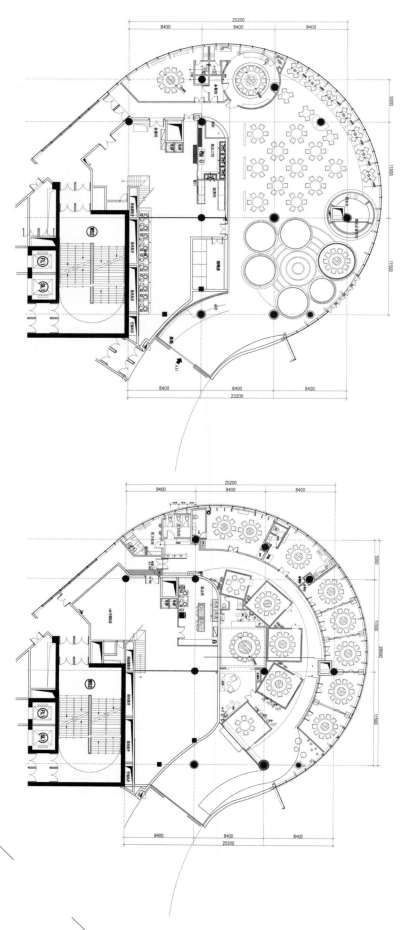

With the inspiration from peonies both of amazing grace and wealthy as well as elegant, this case is another innovation from the design of chain store "Pianyi workshop" by Hehe hall, which places on techniques such as sculptures of peony floral prints and implied meaning of peony color, and applies layout, tone, lighting, material qualities etc to present the atmosphere of wealth and rank as well as prosperous.

With golden appearance both of wealthy and meteoric in the space which is 8 meters from the entrance, which had provided rare spatial terms that there permeates visual impact of noble and elegant as well as dignified and beautiful everywhere such as solid wood being carved with 2 dragons playing with a pearl on arc cashier desk, a background of 8 meters' wood obstruction, red enjoyable paints.

With semi-open type private rooms neighboring the entrance, designers had given plane shaping presentation like peony leaves to it to complete semi-enclosed status. Red lanterns shaping droplights are partly hidden and partly visible among dimly discernible gray fine gauze as if dancing pretty images among airy pavilions and pagodas, having made dining space being more interesting and dramatic.

Large seating area for odd will come into view at the moment of being suddenly enlightened along with the guidance of movements, having a capacity of 200 people to have meals at the same time and provided a scarce place for wedding party and large gathering.

There are private rooms with thick and heavy in colors on the second floor, on both sides of the long and narrow arc corridor, one is golden expressions which eulogize the history of Pianyi workshop, and the other is five private rooms which are designed as clues to peony color, the semi-open space of yellow, red, green, white, ink color, which will make the slightly deep corridor become temperament and interest.

许建国

安徽省建筑工业大学环境艺术设计专业学士

进修于中央工艺美术学院室内设计大师研修班

武汉艺术学院设计艺术学硕士

CIID中国建筑学会室内设计分会会员

国家注册高级室内建筑师

中国建筑室内环境艺术专业高级讲师

中国美术家协会合肥分会会员

安徽许建国建筑室内装饰设计有限公司设计主持

Xiang He Bai Nian

祥和百年

设计单位：合肥许建国建筑室内装饰设计有限公司

设 计 师：许建国

项目地点：合肥市蜀山区霍山路86号

建筑面积：1600 m²

主要材料：水曲柳木饰面、芝麻灰花岗岩、小青砖、
　　　　　仿古地砖

设计时间：2011.12

摄 影 师：吴辉

永和九年，岁在癸丑，暮春之初，会于会稽山阴之兰亭，修禊事也。群贤毕至，少长咸集。此地有崇山峻岭，茂林修竹；又有清流激湍，映带左右，引以为流觞曲水，列坐其次。虽无丝竹管弦之盛，一觞一咏，亦足以畅叙幽情。是日也，天朗气清，惠风和畅，仰观宇宙之大，俯察品类之盛，所以游目骋怀，足以极视听之娱，信可乐也。夫人之相与，俯仰一世，或取诸怀抱，悟言一室之内；或因寄所托，放浪形骸之外。虽趣舍万殊，静躁不同，当其欣于所遇，暂得于己，快然自足，不知老之将至；及其所之既倦，情随事迁，感慨系之矣。向之所欣，俯仰之间，已为陈迹，犹不能不以之兴怀。况修短随化，终期于尽。古人云："死生亦大矣。"岂不痛哉！每览昔人兴感之由，若合一契，未尝不临文嗟悼，不能喻之于怀。固知一死生为虚诞，齐彭殇为妄作。后之视今，亦犹今之视昔，悲夫！故列叙时人，录其所述，虽世殊事异，所以兴怀，其致一也。后之览者，亦将有感于斯文。

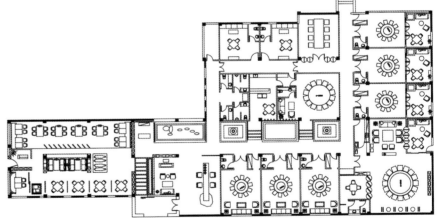

本案以简洁的中式手法，将含蓄内敛与随意自然两种气质完美结合。中式简约风格在设计中穿插运用，不显突兀，反而呈现一种兼容并蓄的美。把所谓古典语汇几何化、图像化、对比化、节奏化等。在线条上化"繁"为"简"；在色调上，讲究稳重而贵气的单一色彩；在空间语言传播上，主张厚重、庄严、质感、奢华，甚至是"开门见山"。

外立面采用中式园林的手法给人眼前一亮的感觉，整体大气，入口则用了照壁的方式，将曲径通幽、移步换景的意境表达了出来，室内也控制得相当得当，整体的节奏以及材料和色彩都把握得很到位，女子十二乐坊的元素运用的也很生动。

　　设计师立足中国本土文化,把徽文化与中式文化完美结合,运用了多种表现形式。设计的某些部分是徽文化的剪辑,设计师借用兰亭序,表现了本案的设计意境。在本案中设计师着力打破设计现状,找到属于中国人自己创新的设计风格,使人真正了解到中国人自己设计中的简约、儒雅的精髓,打造具有诗人情怀的空间。在部分包厢设计当中,就餐区与休息区合理分开,休息区设有很高的天井。

The case perfectly integrates both qualities connotation restraining & casual nature by simply Chinese method. Chinese brief style is applied interspersed in the design and demonstrates beauty that incorporates things of diverse natures instead of showing improperness. Convert the so-called classical words by geometrization, visualizing, comparison or rhythmization. Change "complex" ones into "simple" ones in the aspect of lines, em-phasize on simple color that is dignified and worshipful in the aspect of tone and propose decorousness, texture, luxuriousness or even "coming straight to the point" in the aspect of space language communication.

The Chinese garden style technique used on facades refreshes peoples at their first glances. The dealing result is imposing on the whole. And the entrance applies screen wall facing the entrance and the screen expresses fully the artistic conceptions of winding path's leading to a secluded quite place and landscape changing by step moving. The interior design is under good control also. The whole rhyme, material and color are under full control and elements of Twelve Girls Band are used vividly.

Based on Chinese native culture, the designer perfectly integrates Hui-culture and Chinese culture by using a variety of patterns of manifestation. Some designed parts are edited and arranged Hui-culture. The designer uses Orchid Pavilion to show the design conception of the case. The designer put forth effort to break the design status in this case and find the innovative design style that is suitable for Chinese and enable people understand truly the essences, such as conciseness, briefness and refinedness, within design of Chinese and constructs space that is with poet's feelings. Within the design of some compartments, dining area and relaxing area are reasonably separated and really high courtyard is equipped in the relaxing area.

张健

毕业于杭州某美术类学校

1999年起从事室内设计工作

2007年底创立杭州观堂设计

Little Cook Seafood Restaurant

小厨师海鲜餐厅

设计单位：杭州观堂设计

设 计 师：张健

参与设计：陈蕾

项目地点：杭州保俶路

建筑面积：160 m²

主要材料：旧木板、旧门板、瓦片、石材

摄 影 师：王飞

餐厅以新鲜、美味的特色海鲜为主,在设计中着重运用了海边元素,门头采用大块的石材,营造海边礁石的氛围;一层左侧外墙采用各种瓦片、砖块排列成不同的图案组合,一来外观吸引眼球,二来形成半通透的墙面隔断,路人可以看到室内大厨们忙活的身影,也为厨房的通风提供更好的条件。

　　餐厅室内设计中,采用回收的砖块、门板、窗花、木箱铺就墙面,营造一种随意、放松、粗犷又不失细节的就餐氛围,仿佛海边小镇上的一家小餐厅。餐桌、餐椅、木梁、窗框、隔断、楼梯、扶手也都采用回收的老木板,环保又温馨。

　　由于餐厅本身空间有限,因此在隔断、窗户上面尽量做得通透,临街窗户全部采用透明玻璃,并且尽最大可能的打开,让客人在就餐的同时,能随时与外界新鲜的空气、郁郁葱葱的绿植互动。

　　卫生间与洗手盆的墙面地面则采用破碎的陶瓷片装饰,如咖啡杯、马克杯、陶碗等的碎片,五彩斑斓,形状各异,让客人在如厕同时欣赏不一样的风景。

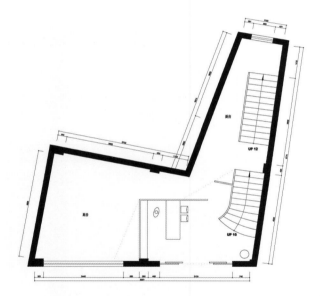

一楼平面布置图 1:50

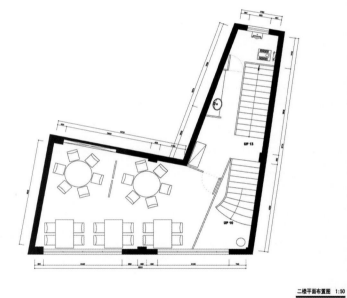

二楼平面布置图 1:50

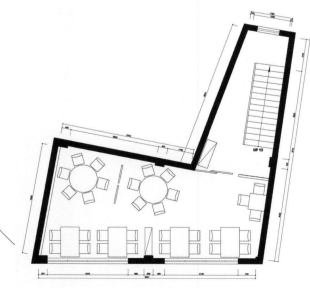

三楼平面布置图 1:50

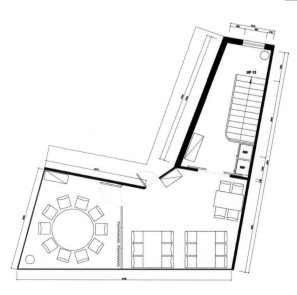

四楼平面布置图 1:50

The restaurant gives priority to fresh and delicious characteristic seafood, applies seaside elements during design, uses massive stone at the top of doors to build the atmosphere of seaside rocks; the left to a bed uses all kinds of tiles and bricks to arrange into different pattern combinations to attract eyeballs by its appearance on one hand, on the other hand, to form semi-permeable wall space obstruction so that not only passerby could see busy figures of chefs indoors but also provide better conditions for airy kitchen.

During the indoor design to the restaurant, uses recyclable bricks, door sheets, paper-cut for window decoration, wooden box to pave wall space to build repast atmosphere of being at will, relaxed, rough without losing details as if a little restaurant on seaside of a town. Dining-table, dining chairs, beams, window frame, obstructions, stairs, handrails are all adopted old board, which are both environmental protective and sweet.

Due to the limited space of the restaurant itself that make both obstructions and windows being permeable to the greatest extent, whole use transparent glasses in the frontage windows and could open at the utmost to enable guests to interact both with fresh air outside and luxuriantly green plants at all times at the same time of repast.

Use crush ceramic chips which are both colorful and different shaped such as coffee cups, mugs and clay bowls to decorate both wall space and ground both of rest rooms and basins, and enable guests to appreciate diverse scenery at rest room.

潘开富

毕业于南京林业大学

南京智点室内设计顾问有限公司设计总监

Tongqing Building Luchou Fu

同庆楼庐州府

设计单位：南京智点室内设计顾问有限公司

设 计 师：潘开富

项目地点：合肥

建筑面积：20000 m²

摄 影 师：吴辉

同庆楼庐州府坐落在合肥市马鞍山路与东流路交口，大厅入口处景观，高梁直柱，徽派之风彰显大气。酒店以徽派园林设计风格为主，或时尚典雅，或恢宏华丽。宴会大厅气派非凡，可承接近200桌各类宴席，配备顶级音响设备、LED显示屏，甚至可以根据顾客需要任意分割空间。在硬件装修选材上用的是进口名贵石材和名师设计的灯具，随便一件便数万甚至百万，为顾客打造最舒适最有质感的就餐空间。

Tongqing Building Luchou Fu is located in the cross of Maanshan Road and Dongliu Road of Hefei City. At the entrance of hall, you can see high beam and right cylinder, style of Huipai and elegance. The hotel is characterized by garden design style of Huipai, fashionable elegance or magnificent and gorgeous style. The banquet hall is extraordinary and can accommodate about 200 tables of all kinds feast, equip with top-level audio equipment, LED display, and can even divide space arbitrarily according to customer needs. As for hardware decoration materials, it selects imported precious stone and lamps and lanterns designed by famous designer. A simple decoration article is with ten thousands or even several millions in order to create the most comfortable and most textured dining space for customers.

宋国梁

中国最具影响力室内设计师

中国百名优秀室内建筑师

ICIAD 国际室内建筑师与设计师理事会宁波区理事

IFI 国际室内建筑师设计师联盟会员

CIID 中国建筑学会室内设计分会会员

新加坡 V.特锐建设集团/总设计师

Yongxiang Fashionable Restaurant

涌香格调时尚餐厅

设计单位:新加坡 V.特锐建设集团

设 计 师:David/宋国梁

参与设计:M.G 设计组、N.O 设计组、纳海川配饰组

项目地点:宁波市鄞州区樟溪北路88号

建筑面积:2300 m²

主要材料:木皮、墙纸、镜面不锈钢、大理石

摄 影 师:刘鹰

撰　　文:野风

涌香格调时尚餐厅地处中国宁波美丽的奉化江畔，位于商业热土舟宿夜江中心地块，环境幽静恬淡。精彩纷呈的时装发布会成为设计师创作的灵感，流动的色彩、混搭的文化元素与本案有着异曲同工之妙，品味、气度、涵养如描绘年轻绅士般写满了从容。餐厅融合现代、西式复古、时尚元素，无论北欧风情、田园诗意、现代情愫，都显得巧妙流畅。整个餐厅洋溢着浓浓的文化气息、书卷满廊、原创前卫艺术装置、手工陶盘、名师画作，从内到外透着优雅，造就了餐厅的特有气质。从一到三楼贯穿的共享空间，为原有楼梯间改造，斜洒阳光，清流潺潺，灵动、活泼、充满生机。包厢风格迥异，各具特色。三层多功能户外餐吧，坐拥甬城绝美夜色，赏佳景、享受美味佳肴、浪漫情怀，为甬城独一无二的特色场所，也是人们沙龙、party的理想去处。不管是家具、灯具、特色工艺装饰，每一个细节都经过设计师精雕细琢，设计师本人创作的装置艺术给空间增添了独特的魅力。

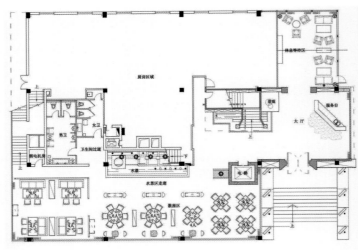

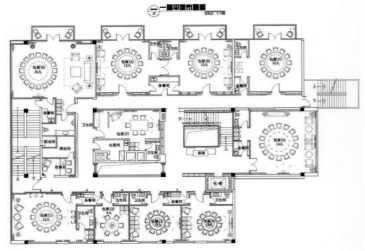

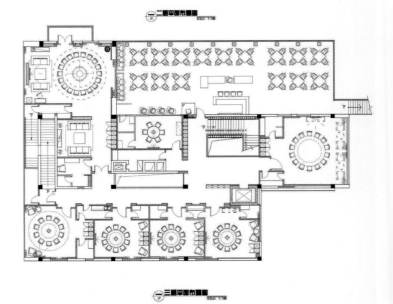

By the side of beautiful Fenghua River, Ningbo, China and being located in the central commercial place, Zhousu Yejiang, Yongxiang Fashionable Restaurant is with quiet and tranquil environment. Colorful fashion show, whose flow of color and mix of cultural elements are different from this design in approaches but equally satisfactory in results, became designers' creative inspiration. The restaurant's demonstration of good taste, bearing and cultivation is shown as those of young unhurried gentlemen. The restaurant integrates modern style, Western vintage style and fashionable elements. No matter it is Nordic style, pastoral poetry style or modern sincerity, the atmosphere is created elaborately and smoothly. The whole restaurant runs over with strong cultural atmosphere; a gallery will full books, original avant-garde art devices, handmade pottery plates and masters' paintings create the unique features of the restaurant and demonstrate a sense of elegance from the interior to the exterior. The shared space connecting the first, the second and the third floors is transformed from original staircase. Oblique sunshine shines, clean water murmurs and creates active and vital atmosphere in the space. And compartments of the restaurant are in different styles and with various features. Three floors of multifunctional outdoor bars provide excellent places for appreciating beautiful night views of Ningbo. That good views, delicious food and romantic themes may be available here makes Yongxiang Restaurant an unmatched characteristic place in Ningbo and also an ideal place for salon and party. Every detail, no matter it is the detail on furniture, on lamp, or on specific artistic decoration, it has been elaborately refined by designers; the art installation created by the designer in person adds unique charm to the space.

周易

1959年 出生于台中市
1979年 以自学方式学习建筑与室内设计
1989年 创设JOY室内设计工作室
1995年 创设JOY概念建筑工作室

Co-exist & Harmonious

客家本色大里店

设计单位:周易设计工作室
设 计 师:周易
项目地点:台湾台中市
建筑面积:895 m²
主要材料:铁件、清水模、枕木、玻璃、大理石、抛
　　　　　光石英砖、竹子
竣工时间:2011.09
摄 影 师:何风摄影

这栋上、下仅两层楼高的独立建筑，清水模的质朴、陶缸水景的写意，加上玻璃帷幕的通透感，在鲜明的建筑语言里，同时注入了业主期许的自然、休息概念，让人第一眼就留下生动印象。

走过户外粗犷的枕木栈道，单侧有着格栅窗花图腾的长列方形步道灯，以典雅的光影诉说着迎宾的热忱。栈道与主建物间规划水景区，以黑色石材砌成的无边界水池里，点缀着四座巨大黑色陶缸，以及状似漂浮的烛台灯与光束涌泉，水池中央一座强调极简线条美学的清水模结构，让室内外有了适度的衔接与屏障，设计者借由这些源于自然界的木、石、光、水等元素，传递一种人造工艺与自然共生的极致，提炼宁静与安逸的环境和谐之美，让所有到访者都能以最放松的心情入内用餐。

一进门所见即为气宇非凡的柜台区，善用建筑物局部挑高逾7m的优势，在柜台后靠端景处以工序繁复的白底浮雕隶书；加上分毫不差的铅字排印技巧，将李白一首潇洒、豪情万丈的《将进酒》，在精妙的聚焦灯光下，进化为气势万千的立体视觉艺术，而点缀文字群间；大小不一却同样别致的红色落款，灵感来自乾隆皇帝喜欢在钟情的书画上捺印，隐喻专属专有的独特性，整个画面的经营，除了意蕴其中的人文涵养，更是一次独到美学的具体实践。此外，文字端景墙前精选黑色石材打造柜台主结构，立面以错落的实木块传递不经修饰的自然感，上方一排利落铁架运用纤细的钢骨深入天花板内强化支撑，架上一长排烛灯是室内外共通的元素之一，综合以上极度情境的构图手法，千年的文化重量与休息的丰富内容尽在其中。

餐厅内部以时尚的黑白对比为主调，但造型语汇上却穿插了许多古老

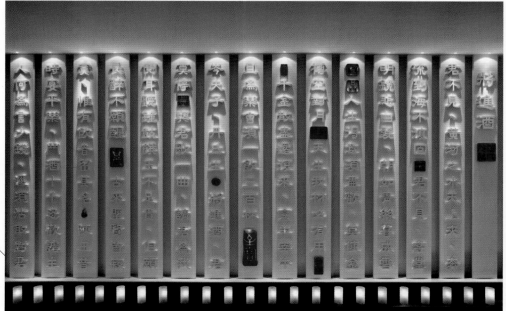

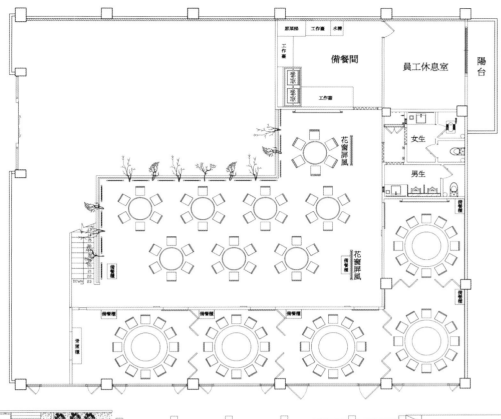

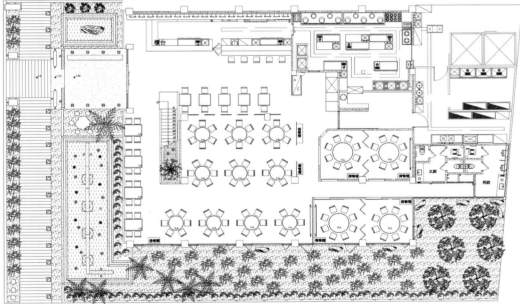

中国的片段,例如漆黑的梯间下缘,一方镜面池子仿佛是户外水景的复刻版,池中一棵全白枯枝,既有白山黑水的泼墨意境,更有日式枯山水的耐人寻味,尤其经过设计师周易一向擅长的灯光烘托,更将生活的无限美好浓缩于眼前的方寸之间。一楼主要为开放的用餐空间,其中一侧运用"有景借景,无景则避"的技巧,将落地窗外优美的竹林景致吸引入内,大大提振食欲和情绪的感染力,不容错过的还有点缀在窗畔白墙与特定包厢内的大幅书法艺术,全是当代知名书法家李峰的作品,其中一幅名为"如易",很巧合地将设计者与业主名字中共有的"易"字带进来,营造既有象征性又意义深远的客制化艺术。另一侧包含夹层上下均使用大量中式窗花分段界定,全数喷白的线条格外立体,一字排开的气势营造"数大便是美"的震撼力,二楼天花板处还有黑色枯枝迤逦而出,串连空间处处呼应的设计主题。

私密包厢的设计同样饶富巧思,在隔墙上缘点缀的白色竹篱与灯光阴影,让人联想起"采菊东篱下"的悠闲,以雅致瓶门发想的入口造型,有着中式园林的书香气质,门上的厢名则以知名的客家聚落命名,这也点出了业主源出于此不忘本的初衷。

JOY Interior Design Studio makes good use of original building's congenital condition such as high interior ceiling room and floor to ceiling continuous windows and use an open field of vision and philosophical, and harmonious logical thinking to bring the Suzhou gardens like quiet indistinct of water, bamboo grove and narrow footway from external view into indoor box view, it also integrate modern classic black-white fashion contrast and Eastern culture fashion vocabulary and change and reconcile according to various proportions. Through the conflict and harmony between the two, it writes a whole new comment on the feeling of a pleasant dinner gathering.

Walk by the outdoor rough railroad tie, one side has a long line of square trail light with grid square window totem; elegant light and shadow tell the sincere welcome. Between the footway and the main building, a water view area is planned, in the borderless pool built with black stones, there are four huge black ceramic vats and candle lights and light beam springs which seem like floating on the pool are used to decorate the pol. In the center of the pool, an exposed form structure which stress minimalist aesthetic lines make appropriate connection and barrier between indoor and outdoor. Though the nature elements such as the wood, stone, light and water which were originated from the nature, the designer delivers the ultimate attainment between artificial technics and nature and refine the beauty of environmental harmony through quiet and comfortable environment.

When entering the door, one could immediately see the extraordinary counter area. The advantage of 7 meter high ceiling in some area of the building is properly utilized, in the scene behind the counter white relief official script made with complicated process and almost perfect print typography skills transform the poem made by Li Bai with chic and lofty heroic feeling under the exquisite focus light into the three-dimensional visual art with strange momentum to decorate the words; the same chic red inscription with different size, and the conduct of the whole picture, in addition to implication with human virtue, is an unique and concrete practice of aesthetics.

The interior of the restaurant is using the fashion white-black contrast as the main tone, but the styling vocabulary is interspersed with a number of fragments of ancient China, for example, the lower edge of the dark staircase, a mirror like pond just like a replica of outdoor water views. A white dead branch in the pond has not only the "splash-ink painting" artistic conception of mountain and rivers, it also has the interesting scene of Japanese mountain and water, especially after the designer-Chou Yi who is always good at utilizing lights to decorate and make the goodness of life concentrated the view in front of our

eyes. The first floor is mainly the open dining space, the beautiful scenery of bamboo forest outside the window is introduced into the interior though the French windows which greatly boost the appetite an emotional appeal.

The design of private box is also full of clever thinking, the white bamboo canopy and light shadow decorated on the edge of the wall make us associate with the laid-back feeling in a famous poem. Entrance door style is using the idea of elegant bottle; it has the sophisticated temperament of Chinese-style garden. The private boxes are named after famous Hakka community, it also points out the owner's idea of not forgetting ones origin.

304

高雄

道和室内设计有限公司　总经理/设计总监
资深室内设计师
中国建筑室内装饰协建筑室内设计师
中国建筑学会室内设计分会会员
建筑装饰装修工程师
IAI 国际室内建筑师与设计师理事会 华南
区及福建代表处理事

Impression of Wang Jiang Nan

印象望江南

设计单位：道和设计机构
设 计 师：高雄
参与设计：郭予书
项目地点：宁德市
建筑面积：287 m²
主要材料：白色烤漆波浪板、黑镜、黑钛、枫木饰
　　　　　面板、白色烤漆玻璃、黑白根大理石
竣工时间：2012.08
摄 影 师：周跃东

印象望江南餐厅位于宁德万达广场内,该店的使用面积为287 m²,餐位数为122人位。

本案黑白色相间、简约、线条具有一种直抵人心的力量,整体设计呈现时尚、简洁的气质。

黑白色反差、镜面反射的场景,具有强烈的空间感。店内家具均以黑色皮革及大理石台面为主,墙面造型均以凹凸的波浪板及镜面为主,强烈的厚重感与跳跃简洁的线条给客人带来感观的冲击,也体现了餐厅的品牌定位。整体色彩对比强烈,但空间更富于流动和变化。

白色烤漆玻璃墙面若隐若现的山峦,给人无限遐想。仿佛置身江南水乡。

入口大厅的处理是本案的出彩之处,以玻璃钢雕塑倒挂或坐的各异姿态,加上中式的白色喷漆鸟笼以及栩栩如生的小鸟,时尚现代感与东方气质的碰撞,为整个餐厅增添许多各异的色彩。

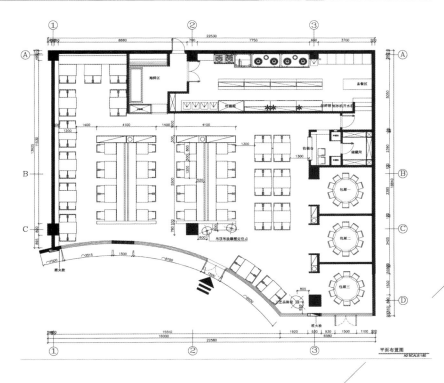

平面布置图

The restaurant of Impression of Wang Jiang Nan is located in Utama shopping center in Ningde, which covers area of 287 square meters and boosts dinning spaces of 122.

Black is alternating with white, being brief, and lines of this case have powers of straight through the heart, whole design is presented on temperament both of fashion and brief.

Contrasting black with white and the scene of mirror reflection have strong sense of space. The furniture in the store is mainly black leather and marble platform, modeling of wall space is mainly concavo-convex wave board and mirror surfaces, strong massive sense and lines both of jumping and brief had brought customers impacts of perception, which will also reflect brand positioning of the restaurant. Contrasting colors of the whole, but spaces are richer both in flow and change. Wall spaces of white baking finish glasses and mountains both of partly hidden and partly visible had brought infinite reveries, as if placing oneself in Yangtze River delta.

The treatment of the entrance to the hall is distinctive, different gestures of sitting or dropping away of sculpture of glasses reinforced plastic, and white spray painted birdcage of Chinese style with lively birds, sense both of fashion and modern crashes eastern temperament, which had added much different color to whole restaurant.

310

潘鸿彬　谢健生

潘鸿彬

泛纳设计事务所创始人

香港理工大学设计学院助理教授

香港室内设计协会副会长

IFI国际室内建筑/设计师联盟执委

谢健生

室内设计师

香港理工大学设计学院美术及设计学荣誉学士

香港室内设计协会专业会员

担任香港生产力培训学院讲师

2005年成为泛纳国际设计有限公司合伙人

Izakaya Singer

圣家居酒屋

设计单位：PANORAMA泛纳设计事务所

设　计　师：潘鸿彬

参与设计：谢健生、黄卓荣

项目地点：深圳

建筑面积：340m²

竣工时间：2012.07

摄　影　师：吴潇峰

"圣家居酒屋"是深圳市新开辟的饮食和娱乐中心"欢乐海岸"的一间日式餐厅,其主要景观的设计给人"家"的感觉,将传统的日本居酒屋餐厅体验提升到新的潮流水准,将收集起来的各种材料循环使用,提高其用途和观赏价值。

餐区一:

    棕色的开放式天花板之下,倾斜的木制屋顶结构在建筑上对传统的日式居酒屋作出了新的诠释。

    在铁板烧台面和长木台的上方悬挂着以空米酒瓶和钨丝灯泡装配成的灯饰。不同风格的白色餐椅与木地板成了对比,使整体温暖的色调产生随意但很一致的图案。

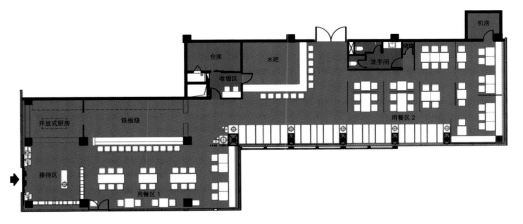

餐区二:

    由天花吊下的绳帘将餐区分隔成不同的空间。各种不同的座位摆设给食客提供2人,4人,6人及8人的餐位,尚有贵宾房以满足客人所需。在富有特色的墙壁挂满了巨大的黑白色日式烹饪器皿与黄芥末色块拼贴成的装饰,使餐厅独具风格和温馨的品牌形象得以突显及使食客胃口大开。用竹筷编织的吊灯和台灯使整个餐区沉浸在温暖舒适的气氛中。

Izakaya Singer is a Japanese restaurant situated at the newly-opened food & entertainment hub "OCT Bay" in Shenzhen, China. A main scene of "Dining Room" was created to give a "home" feeling and move the traditional izakaya's dining experience to a new level of trendiness. Up-cycling of different found objects provide unique casual dining experience for the whole space:

Dining Area 1

Architectural re-interpretation of traditional "Japanese wine house" was given by slanted timber roof structure under brown-painted open ceiling.

Used sake bottles & tungsten light bulbs were re-configured to become pendent lamps above timber bench seatings & Teppanyaki counter respectively.

White dining chairs in various styles set up contrast to the timber floor and provide random yet consistent pattern to the general warm-tone environment.

Dining Area 2

Ceiling suspended rope screens provide flexible compartmentation to the area. Various seating patterns of booth, table of 2 / 4 / 6 / 8 and VIP rooms are introduced to cater for different customer's needs.

Super-sized black & white images of Japanese food cooking utensils collaged with mustard yellow palettes at feature walls create a warm & stylish identity to this new brand and help increased the customer's appetite.

Custom-designed pendent and table lambs made of bamboo chopsticks completed the cozy and warm dining experience.

陈亨寰

来自台湾
大匀国际设计中心协同主持人
台湾中生代设计师
IFI国际室内装饰协会专业会员
中国室内装饰协会专业会员

李巍 Amy Lee

1985年生于中国辽宁省葫芦岛市
毕业于中国上海东华大学装潢艺术专业

Sanya Qixian Ridge Western Restaurant
三亚七仙岭西餐厅

设计单位：大匀国际设计中心
设 计 师：陈亨寰、李巍
软装设计：上海太舍馆贸易有限公司
项目地点：三亚
建筑面积：1013 m^2
竣工时间：2012

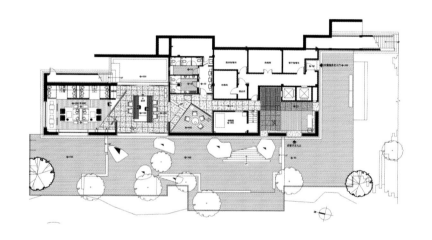

明窗延静书，默坐消尘缘

　　日本茶道宗师千利休曾说过："须知道茶之本，不过是烧水点茶"。

　　空间设计本质亦是通过静虑，从平凡的生活中去契悟设计的大道。

　　我们将这个半开放、半私密、半公共的空间溶于天蓝海碧，山清水秀之中。利用庭院、特色走廊等过渡空间，以达到室内外的精美融合。古朴与现代的巧妙混搭，取材天然与自然相融，使人得以在大自然的环境中更惬意体验当下的放空。

　　从来佳茗似佳人，禅茶一味悟自心。1F茶书吧的空间格局被分割成几个细部，更注重于私密空间与公共空间的互动关系。多层次的过厅更给空间增添了仪式感。此外，大面积的户外景观又与室内空间遥相呼应。于此中，若坐若卧，亦是怡然自得。

　　其下如是，其上亦然。2F西餐厅空间设计格局更是大面积的运用到落地窗，仿若隐逸空谷中一颗璀璨的钻石。延续1F雕花屏风的局部隔断，光影互动交织成趣，给空间更添私密性。以竹木饰面的吊顶区分整体空间。动线的理性规划，让客户在取餐时更方便。宽大舒适的布艺沙发，更是将整个用餐环境带来了更多的轻松氛围。户外的用餐空间，既是日丽风和或月明星朗的逍遥自在。

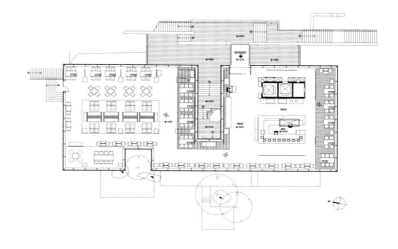

Eading quietly beside the bright window, Sitting silently to eliminate carnal thoughts Sen Rikyu, the master of Japanese tea ceremony, said, "The essence of tea lies in boiling water to make tea".

The essence of space design also lies in understanding the significance of design through meditation and ordinary life.

We blend this semi-open, semi-private and semi-public space with blue sky, azure ocean, green mountain and clear water. We perfectly combine indoor and outdoor space by courtyard and unique corridor. The skillful mix of primitive simplicity and modernism, and the combination of natural materials with the nature en-

ables people to enjoy the more pleasant leisure in the great nature.

The good tea is like a beautiful lady. Therefore, the understanding of Zen and tea is from the heart. Tea and Book House in 1F has been divided into several sections, paying greater attention to interaction of private space with public space. Multi-layer corridor brings a sense of ceremony to the space.In addition, vast outdoor landscape echoes with indoor space from distance, making you comfortable and delightful in sitting or laying down here.

The upstairs is as comfortable as the downstairs. Western Restaurant in 2F has

been decorated with large French window, resembling a brilliant diamond hidden in the valley. The partition with engraved screen in 1F has also been applied here to add more privacy to the space with skillful application of lights and shadows. The bambood and wood ceiling separates the whole space. Dynamic and rational partition enables customers to fetch food easily.The large cozy fabric sofas make the dining environment more relaxed. On the other hand, the outdoor dining space is also pleasant with gentle breeze or romantic with bright stars.

Contents·目录

何宗宪

新加坡大学　建筑学士

香港大学　建筑硕士

现任香港室内设计协会副主席、香港专业教育学院专业顾问。

近期主要荣膺

2003年获选日本Studio Voice亚洲创作VIP之一。

香港设计师学会选录为2021二十名年轻建筑师之一。

香港《Perspective透视》选录为40 under 40大中华区40位设计新人类。

2008年获选为香港传艺节十大杰出设计师。

The Loft Restaurant
The Loft 餐厅

坐落地点：香港东涌

设计公司：*Joey Ho Design* 何宗宪设计

设计师：何宗宪

主要材料：油漆、木地板、墙纸、可丽耐、清玻璃、清镜、布料、人造皮、石材、不锈钢

摄影：*Mr. Dick*

本案设计的出发点是结合进餐体验和用餐地点，希望透过餐厅的室内环境演绎出饮食的新文化意识，把意大利的饮食文化融入本地，让客人细嚼享用食物本身之余，也品尝室内空间的情怀和味道。设计师的灵感主要来自原有的建筑结构，以美国纽约的废弃工厂改造屋(Loft)作为创作蓝本，阁楼设计巧妙地把本来的空间营造成"屋"的环境，风格独特，在宽敞和开放的大厅散发轻松、温馨的居家气氛，为客人带来犹如在家一般的用餐享受。

The starting point is the case with the design dining experience and dining venues, indoor environment, the restaurant aims to deduce the diet conscious culture, the Italian food culture into the local, ponder guests enjoy the food itself, but also an indoor tasting feelings of space and taste. The designer's inspiration mainly from the original building structure, to transform the United States abandoned factory in New York House (Loft) as a creative blueprint that should have been cleverly designed loft space to create a "house" environment, unique style, the spacious and easily distribute the hall open, warm home atmosphere for the guests to bring like dining at home to enjoy.

本案设计干净利落，简约的阁楼设计布局提供充足的光线和宽敞的空间，营造出无加工但又时髦别致的格调。设计师在餐厅内创造视觉层次，在原有的建筑空间内加入空间，用上框架设计，增加整个环境的穿透性，提高空间的使用效率之余又能建构室内的层次。再者，餐厅设计素材采用原材料，例如木材地板和墙身的红白砖块，带出一种不经修饰和陈旧的味道。天花和框架结构的红白油漆形成，但色调同时又和墙身的红砖互相呼应，而且物料与原材料形成简单对比，新旧元素并存。

Case design neat, minimalist loft design layout to provide adequate light and spacious, creating a non-processed but funky style. Designers create visual hierarchy in the restaurant, the original building space to add space, and use frame design, to increase the penetration of the environment, improve the efficient use of space than can construct the indoor level. Furthermore, the restaurant design raw material used, with a kind of modification, and without the old flavor. The red and white ceiling paint and frame structure formation, but also tones and red brick walls to each other, and the materials and raw materials to form a simple comparison.

走进餐厅的另一面，斜顶的天花是原本的建筑结构，设计师善于采用这独特位置的空间结构，把阁楼的概念延续，完美地应用在餐厅这一个角落，配以充足的光线和和暖的色调，还有和谐舒适的家庭布局，为这阁楼创造富时代感的魅力。绘画艺术是本区的重点，设计师特意度身设计墙壁和天花的插图，以居家环境作构图意念，包围整个餐区，为冷冽的墙身增添一丝暖意和亲切感，在空间添上个性和惊喜，散发浓厚的温馨气息。居家绘图覆盖全场的布局不但为阁楼餐区带来热情款待的轻松气氛，空间更能培养客人的归属感。

Walking into a restaurant the other side, slant ceiling is the original building structure, the designers are good at using the spatial structure of this unique location, the attic ("Loft") the concept of continuity, the perfect application of this corner in the restaurant, accompanied by adequate light and warm colors, and layout of a harmonious and comfortable home for the attic to create this contemporary charm. The art of painting is the focus of this area, designer walls and ceilings specially tailor-made illustrations, to the home environment for composition ideas, surrounded by the dining area, as the cold wall to add a trace of warmth and intimacy, the space character added and surprise, distributing strong warm atmosphere. The layout of the home drawing the audience not only covered dining area for the attic to bring the relaxed atmosphere of warm hospitality, guests space to better develop a sense of belonging. The case of different shapes, colors match not only depicts the unique personality traits, and successfully created the appropriate restaurant, fashion brand names, activate the senses and taste buds of the guests.

程绍正韬

逢甲大学建筑系讲师

东海大学建筑系讲师

真工建筑设计公司 总设计师

真品空间艺术股份有限公司 艺术指导

明建筑艺术细装工程股份有限公司 艺术建筑师

联合建筑规划股份有限公司 环境设计师

近期主要作品

2008 台中生活空间奖 TEA WORK 人水私房

2008 园治奖 UFO软体科学经贸园区公共空间暨都市设计

2008 都市景观评选大奖 UFO软体科学经贸园区公共空间暨都市设计

2008 高捷大寮站公共艺术第一名

2008 室内TID奖 "市政观邸"公共空间景观暨建筑设计

2008 室内TID奖 "私房泰"顶级泰式料理餐厅

2008 室内TID金奖 "一即一切"

2009 室内TID金奖 "House-M"

2010 亚太空间设计师协会 APSDA Grand Award 金奖 "House-M"

2010 台湾创意设计中心 TDC 室内金点设计奖 "House-M"

2010 第8届台湾设计博览会 年度主题区 "善念"

2011 台北世界设计大展"国际室内设计展"主题 "影响与汇合"

2011 台湾当代设计师联展 主题"交锋"

Private Thai Restaurant
私房泰餐厅

项目地点：中国台北市庆城街1号2楼

面积：*1220 m²*

设计师：程绍正韬

参与设计：张嘉伦、王俪蓉

主要材料：洞石、石皮、原木、树皮、竹皮、毛竹、漆艺铁艺、玻璃、丝绒、壁纸、手工艺术创作

摄影师：李国民

忙忙碌碌的现代生活中，平凡、简单而安宁和谐有时反而成为一种奢求。本案为位于中国台北私房泰泰式餐厅，这里面对如流水般的人潮，餐厅企图为每日需要依赖地铁的普通大众提供一个闹中取静的场所。

Busy modern life, ordinary, simple, peaceful and harmonious life can sometimes become a luxury. The case is located in Chinese Taipei Private Taytay restaurant, where the face like water as the crowd, the restaurant attempted to rely on the subway every day to provide the general public a quiet place.

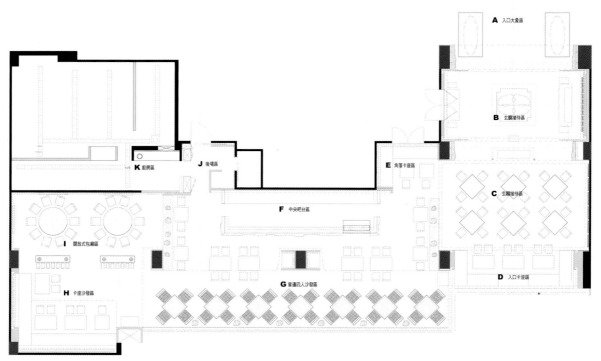

平面图/Floor plan

设计师首先在空间中布置了8个机能区域，除了厨房、吧台等机能区域之外，所有的服务区域被以环境心理学的模式建构，每个区域形成的景观餐饮小区都赋予一种独特的自我风格，再来是材料的细部设计。也就是说：先用形状构成所谓静观城市的"窗口"，在一个纷扰的世界去为人们建造一个属于他们自己的身心避风港。

Designers first layout in the space of eight functional areas, in addition to the kitchen, bar and other functional areas, all of the service area is in environmental psychology model construction, food and beverage landscape in each region forming a district is given a unique personal style, again the material of the detail design. In other words: the first with the city's wait and see the shape constitutes the so-called "window" in a world of troubles for people to build a mind of their own haven.

对于当代高品质餐饮空间的设计师来说，消费者的消费心理要优先于物理机能来被考虑，特别是在一个这样的交通通道与商业运作核心中。用餐当然就绝不只是一种纯粹的生理行为，用餐已经是一种心理学。在这样的一个文明区块，每当用餐时刻，甚至其他时段，人们思考的除了美食之外；更多的时候已经是在思考寻找一个完全放松的自己的空间。

For high-quality contemporary dining space, designers, consumer psychology to give priority to physical function to be considered, especially in such a traffic channel with the core business operations. Of course, not just eating a purely physical act, is a psychological eating. Block in such a civilization, when dining time, and even other times, in addition to food than people think; more often already thinking about looking for a complete relaxation of your own space.

于是，谨慎而细腻的环境心理学式的思维模式便被应用。各个风韵相异的用餐区域被非常谨慎地使用，每种材质与色泽被深化；秀雅竹材、原野味风化樟木柱、极粗犷有力的黝黑巨大石块、细致华贵的艳丽丝绒还有那些让人还原性灵纯粹的细腻白色涂料。它们之间被巧妙重组，在复杂的肌理系统变化中营构了许多不同风格，却都展现一种自然优雅风韵的格调。

Thus, the careful and delicate environmental psychology of thinking style will be applied. Each charm is distinct dining area very carefully used, each material and color is deepened; bamboo, camphor wood columns, a very powerful dark rugged rock, fine luxurious velvet still delicate white paint. Between them are subtle reorganization, changes in the complex texture in the business structure of the system in many different styles, but all show the elegant charm of a natural style.

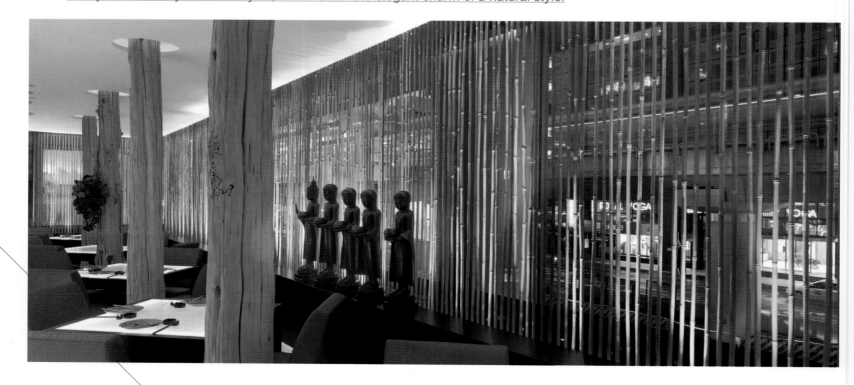

蒋建宇

宁波宁海人，现居杭州。

2001年，组建大相艺术设计公司

2011年，组建大相莲花陈设艺术公司

Shanghai New Rongji

上海新荣记

项目地点：上海市卢湾区人民广场

建筑面积：*1500 m²*

设计单位：杭州大相艺术设计有限公司

设计师：蒋建宇、李水、董元军、郑小华、虞华明

主要材料：珍珠黑、黄古铜、银龙灰大理石、木地板、柚木饰面

摄影师：贾方

他绝不是一般意义上豪华舒适的餐厅，而是在精神上给你安慰，在味蕾上让你惊讶的一个场所，他有着细微的前卫，也有着商业上所需的所有元素，但是他细密的传达了一种亲密。上海新荣记可说是具有一种冲突中取得谐和的最佳平衡。

He is not an ordinary sense of the luxurious restaurant, but in the spirit to give you comfort, you surprised at the taste buds of a place, he has this subtle avant-garde, but also has all the elements needed business, but he fine communication of an intimate. Shanghai New Wing Kee can be said to have a conflict in harmony to achieve the best balance.

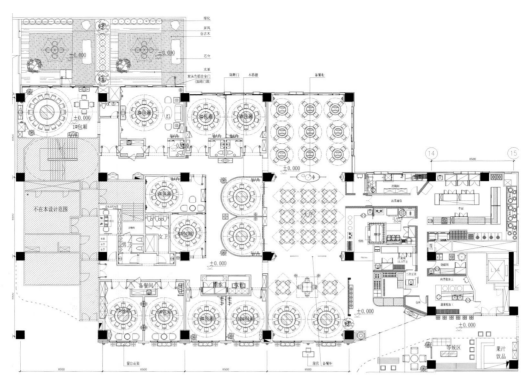

平面图/Floor plan

上海新荣记以人本位的观念来设计，超越单纯追求视觉的设计观念，此案本着可能调动一切感官来感受着饕餮大餐，即眼，舍，耳于一体。正因为如此，他们更能回归到酒店的基本概念，一个全身体验过程诞生。

The restaurant's design to people beyond the pursuit of visual design with a simple concept, the case may be to mobilize all the senses to the spirit of feeling good dinner, eye, homes, ear. Because of this, they better return to the basic concepts of the hotel, a body to experience the birth process.

奢华是相对的，对一个超繁华的都市，相对浮躁的内心，相对纷杂的世界来说，奢华则应是安宁与沉静。按照这个出发点，酒店希望能在喧闹的城市中，逃避上海永无止境的匆忙，希望能让客人在进入酒店时在心情上有巨大的转变，嘈杂与繁盛逐渐蜕变成宁静与安详。

Luxury is relative, of a super-prosperous city, relatively impetuous heart, relatively confused world, should be peace and quiet luxury. According to this starting point, the hotel hopes to noisy cities, Shanghai to escape the never-ending rush, want to make customers feel when entering the hotel there is a huge change, gradually transformed noisy and busy to quiet and serenity.

李泷

中国建筑学会室内设计分会 会员

IAI亚太建筑师与室内设计师联盟 理事会员

宽品设计顾问有限公司 设计总监

近期主要作品

北京山水文园会所

佳科集团（中国）总部

创冠集团（香港）总部

鼓浪屿那宅精品酒店

中石化福建总部大楼

长沙优山美地

北京万城公馆样板房

冠豸山温泉度假酒店

观音山国际商务营运中心

Bafang Zhuan Restaurant

八方馔养生餐厅

项目面积：**900m²**

项目地点：福建 厦门市

设计公司：厦门宽品设计顾问有限公司

主案设计：李泷

参与设计：张坚 林惠平

八方馔是一间以制作养生美食拥有八大菜系为特色的餐厅，整个空间注重营造与料理相互交融的氛围，餐厅设计以简约中式为基调，整体色调以米色为主题色，搭配灰色作为过渡，使整个空间沉稳低调并不失时尚。

Bafang Zhuan Restaurant is one for the production of health food has eight restaurants featuring cuisine, focusing on the entire space to create a blend between the atmosphere and cuisine, the restaurant design to simple Chinese as the keynote theme of the overall tone of beige color, with gray as a transition, so that the whole space is not lost steady low-key fashion.

外观黑色格栅由"八"字提炼造型，美观且耐人寻味。大堂富有中国风的黑金色工笔漆画，简洁丰富的肌理墙面，素雅禅意的布艺隔断，写意花鸟的麻纸吊灯无一不渗透着浓浓的东方意韵。二楼以一片鸟语花香的景象拉开序幕，简洁明亮的餐区，散发着浓郁古典气息的水墨意向纵横交错，营造空间丰富层次并塑造神秘感。

Appearance of the black grille from the "eight" shape refined, beautiful and intriguing. Lobby full of Chinese style black and gold lacquer brushwork, rich texture walls simple, elegant Zen cloth partition, freehand flowers of hemp paper chandeliers are all permeated with a strong oriental Prose. On the second floor with a fine spring scene kicked off, simple bright dining area, exudes strong intention to criss-cross the classical atmosphere of ink, to create space and create a rich level of mystery.

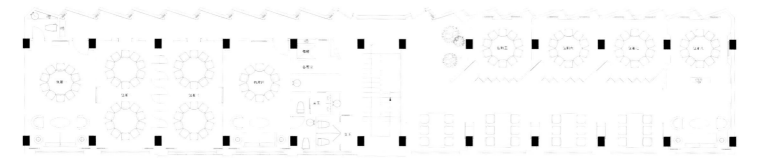

平面图/Floor plan

整体空间将中式古典元素巧妙地融入现代空间中，色彩，材质，造型和谐共存，极具美感的视觉效果共襄心灵盛宴。

Chinese classical elements of the overall space cleverly integrated into the modern space, color, texture, shape harmony, great beauty of the visual effects grand spiritual feast.

刘旭

成都蜂鸟设计顾问有限公司 主设计师

2008年，部分设计作品分别收录在《中国室内设计年鉴之商业空间》、《东方风情》、《中国顶尖样板间》、《中国室内设计年鉴之特色餐饮》

2009年，部分设计作品收录在畅销海外的《室内公共空间》

近期主要作品

2008 年，重庆顺风123西城天街店、三峡广场店室内空间设计；

2009 年,重庆顺风123大都会店室内空间设计；

2009 年,重庆顺风123北京会所建筑外观方案及室内空间设计；

2009 年,天府金茂房产光明城市一、二期样板间的设计；

2010 年，重庆顺风123湖景会所一、二期建筑外观方案及室内空间设计；

2010年，成都鼎月美食会所建筑外观方案及室内空间设计；

2010年，重庆瑞泰丰成都龙湖三千集店室内空间设计；

2010年，金海华集团旗下大城小爱（餐厅）室内空间设计；

2010年，天津新煮意万达店室内空间设计；

2010年，湖南株洲红花树•湘江一号会所建筑外观设计与室内空间设计

Shunfeng 123 Beijing Club

顺风123北京会所

项目面积：*1600m²*

项目地点：北京朝阳区工人体育场西门

设计公司：成都蜂鸟设计顾问有限公司

主要材料：石材、*PU*硬包、墙纸、小牛皮、饰面板

摄影：唐嘉骏

消火栓

本案，以"予方于圆"的寓意为灵感来源。整体空间以延续建筑和平面设计的黑白色调搭配为主，以直线为骨架的平面规划，加以时尚雅致的家具搭配及陈设，增强空间延伸感的同时，缔造出一种新鲜的低调与奢华。

入口纯黑色的拱形门明晰地将室内与外界分隔，传递出一种踏进门，就恍若步入一个时尚、优雅，具有浓厚人文气息的空间。接待区墙面透光方格的渐变造型看似漫不经心的组合，实则出奇制胜，与中式简化造型接待台上的蝴蝶兰，顶上的复古金箔饰面相呼应，呈现出若隐若现的禅意，在演绎新式中国道家文化的同时，为接待厅注入了具有东方美的时代感。

Project to "to square the circle," the implication for the source of inspiration. The whole space to continue building and graphic design with the main black and white, straight-line floor plan for the skeleton, to be stylish with elegant furniture and enhance a sense of spatial extension, create a kind of fresh, low-key and luxurious.

Entrance, arched black interior with the outside world goalkeeper separated and out, entered the door, into a stylish, elegant, with a strong cultural atmosphere of the space. Reception area, Chinese simplified style reception stage orchid, top retro gold finishes echoes, showing a looming Zen space into a beautiful contemporary oriental.

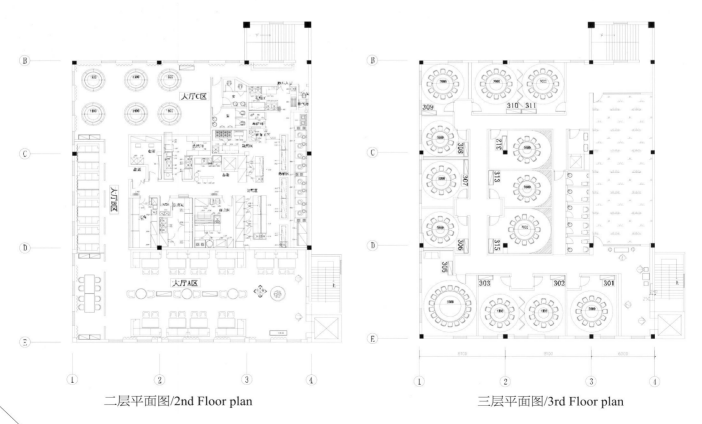

二层平面图/2nd Floor plan　　　　　三层平面图/3rd Floor plan

卡座区墙面为与大厅相呼应，特以橡木边框配以做点缀的浅色小牛皮毛饰面，使沉稳的色调中有了新的突破，室内夸张的黑色球形金属铜质吊灯与整个风格拉开了极大的反差，为空间注入新的血液。

Deck area wall of the hall echoed with, especially to make decorative oak frame coupled with light-colored calf fur finishes, the calm tone of a new breakthrough in indoor exaggerated spherical metallic copper chandeliers and black the style opened a great contrast, for the space to inject new blood.

包房通道两侧以PU硬包饰面，与墙面构成节奏相同的方形壁灯上下发光，在略显深邃狭长的通道里，若隐若现地烘染出清静无为的道家思想；包房内利用现代的手法和材料，运用不同的元素来表现空间的设计主题：整个区域内以简洁明朗的弧形线条为主，避免了直角带给人的生硬与抵触，顶上的圆形吊顶也与之呼应，为了将新东方风格极具代表的圆与方的结构于一体，部分包房内以中国传统象征吉祥如意的玉环做装饰，墙面大多选用色调高明度的几何结构组合墙纸饰面为主，地面以具有内敛不失稳重之感的深色木地板斜拼铺装，几种不同材质组合在一起产生强烈的视觉碰撞，让统一于内在的现代中式元素，在室内外空间自由转换和穿插。

Rooms on either side with PU hard pack finishes, with the same rhythm form the wall and down the square wall light, slightly deep in the narrow channel, the looming drying dyed quietism of Taoism; bag room used modern techniques and materials, the use of different elements to the performance space design theme: simple and clear throughout the region to the curved lines, avoiding the rigid right angles to bring people and conflict, with the top of the circular ceiling is also echoed Oriental style to very representative of the new circle and the square of the structure in one part of the package room to the traditional Chinese symbol of good luck in Yuhuan decoration, the wall most of the smarter choice of color combination of the geometry-based wallpaper finishes, floor to have a sense of restrained yet stable inclined to fight the dark wood floor coverings, a combination of several different materials together to produce a strong visual impact, so that a unified modern Chinese elements in the inner, freely switch in indoor and outdoor space and interspersed.

刘旭

成都蜂鸟设计顾问有限公司 主设计师

2008年，部分设计作品分别收录在《中国室内设计年鉴之商业空间》、《东方风情》、《中国顶尖样板间》、《中国室内设计年鉴之特色餐饮》

2009年，部分设计作品收录在畅销海外的《室内公共空间》

近期主要作品

2008 年，重庆顺风123西城天街店、三峡广场店室内空间设计；

2009 年，重庆顺风123大都会店室内空间设计；

2009 年，重庆顺风123北京会所建筑外观方案及室内空间设计；

2009 年，天府金茂房产光明城市一、二期样板间的设计；

2010 年，重庆顺风123湖景会所一、二期建筑外观方案及室内空间设计；

2010年，成都鼎月美食会所建筑外观方案及室内空间设计；

2010年，重庆瑞泰丰成都龙湖三千集店室内空间设计；

2010年，金海华集团旗下大城小爱（餐厅）室内空间设计；

2010年，天津新煮意万达店室内空间设计；

2010年，湖南株洲红花树•湘江一号会所建筑外观设计与室内空间设计

Shunfeng 123 Lake Club II
顺风123湖景二号会所

项目面积：*1500m²*

项目地点：重庆

设计公司：成都蜂鸟设计顾问有限公司

主要材料：*PU皮料、橡木、仿古砖、透光云石*

摄影：唐嘉骏

本会所定位高端，就餐区域均以包厢的形式呈现，基于此前提，设计师希望能将极具私密性的包厢打造成一处静幽，宛如与世隔绝般的清雅空间。因此，在创作中故意规避太多繁杂的装饰构成与造型，而将更多的心思转移到对细节的处理，引入自然光源、配合典雅、自然的装饰手法将室外与室内完美地结合在一起，整个空间都透露出一种内敛沉稳的人文内涵。

The positioning of the high-end clubs, dining area are presented in the form box, based on this premise, the designers hoping to play a very private nature of the box caused a quiet secluded, like isolated like elegant space. Therefore, to avoid too much deliberate creation of complicated decorative form and shape, and put more thought into the details of the transfer process, the introduction of natural light, with elegant and natural way to the outdoor and indoor decoration combines, the entire space revealed a restrained calm the cultural connotations.

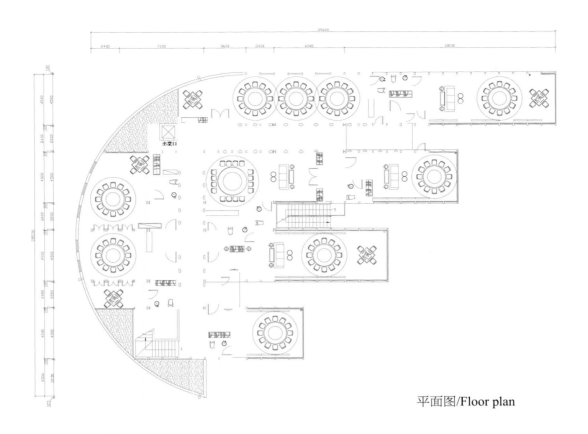

平面图/Floor plan

通道处采用了看似无序实而有序地排列方式，不规则地布以透光云石做软性划分，达到具有异曲同工之妙的隔断效果，包房内选取了象征文人气质的中式家私与富有禅意韵味的台灯、尔雅隽永的壁画相互辉映，体现了中国传统文化的精髓，真正实现了人与自然真实和谐的零距离接触。

Channel at the use of a seemingly disordered solid and orderly arrangement, irregular light cloth to marble, to cut off the effect, select the package room full of Zen flavor of lamps, murals reflect each other, reflecting the Chinese essence of traditional culture, truly a real harmony between man and nature contact.

本案，设计师将重点放在实现空间的稳定与有效性，展现温暖洗练的生活态度。以流畅的空间布局和细部摆设，融合"新亚洲主义"的设计理念，为空间注入一种含蓄清雅的气质，让来此就餐的顾客由内而外的体会到一种与喧嚣尘世相隔的清幽脱俗。

Designers focus on achieving stability and effectiveness of space to show warmth succinct approach to life. To smooth the spatial layout and detail display, integrated "new Asianism" design concept for the space into a subtle and elegant temperament, so that the customers come from the inside out dining experience to a quiet and bustle of the earth separated and refined.

刘旭

成都蜂鸟设计顾问有限公司 主设计师

2008年，部分设计作品分别收录在《中国室内设计年鉴之商业空间》、《东方风情》、《中国顶尖样板间》、《中国室内设计年鉴之特色餐饮》

2009年，部分设计作品收录在畅销海外的《室内公共空间》

近期主要作品

2008 年，重庆顺风123西城天街店、三峡广场店室内空间设计；

2009 年，重庆顺风123大都会店室内空间设计；

2009 年，重庆顺风123北京会所建筑外观方案及室内空间设计；

2009 年，天府金茂房产光明城市一、二期样板间的设计；

2010 年，重庆顺风123湖景会所一、二期建筑外观方案及室内空间设计；

2010年，成都鼎月美食会所建筑外观方案及室内空间设计；

2010年，重庆瑞泰丰成都龙湖三千集店室内空间设计；

2010年，金海华集团旗下大城小爱（餐厅）室内空间设计；

2010年，天津新煮意万达店室内空间设计；

2010年，湖南株洲红花树•湘江一号会所建筑外观设计与室内空间设计

Shunfeng 123 Lake Club III

顺风123湖景三号会所

项目面积：*1500m²*

项目地点：重庆

设计公司：成都蜂鸟设计顾问有限公司

主要材料：爵士白石材、铝板、小牛皮毛、PU皮料、橡木

摄影：唐嘉骏

本案空间划分主要分为接待区、就餐区、包厢区，设计以西方时尚、贵族的文化背景为主题思想，以寻求现代英伦风格的延续和创新为理念，以具有经典色之称的黑、白色为主色调……我们力求通过设计创作——功能划分、动线引导、细节装饰、氛围营造，将会所打造成一件空间艺术品。

This case space is divided into the main reception area, dining area, balcony area, designed to Western fashion, the theme of the cultural background of noble ideas, to seek a continuation of the modern British style and innovation as a concept, known as a classic black color, white color ... we strive through creative design - functional partitioning, dynamic ray guidance, details of the decoration, to create an atmosphere, the play will result in a space of art.

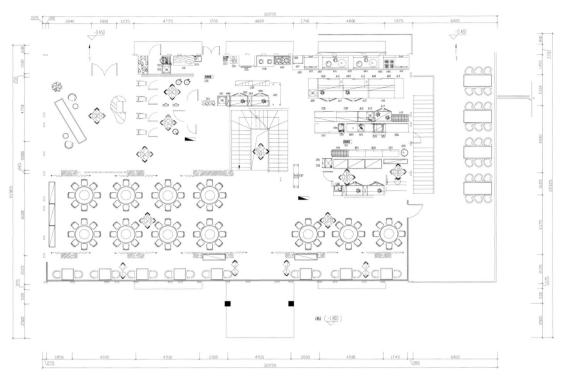

一层平面图/1st Floor plan

大厅在追求实用功能的前提下，降低了对装饰和形势的过分追求，力求在功能与形式之间达到新的平衡。稳重大气的意大利进口爵士白大理石与墙纸细腻柔媚的质感形成生动的对比；价格不菲的ARTEMIS壁灯、FEND的新款坐椅都为整个空间注入了一股高贵的气质；餐厅引入的自然光源，配合典雅、自然的装饰手法将室内外完美地结合在一起，整个空间都透露出一种内敛沉稳的人文内涵。

Hall in the pursuit of practical functions under the premise of reducing the decorative and the situation of over-pursuit, and strive to achieve the function and form of the new balance. Stable atmosphere with imported Italian white marble wallpaper jazz delicate and lovely texture to form a vivid contrast; expensive ARTEMIS wall, FEND of the new chairs are for the entire space into a wave of noble qualities; restaurant to introduce natural light, with elegant and natural way to decorate the perfect combination of indoor and outdoor, the whole space revealed a restrained calm the cultural connotations.

为达到整体与细节的统一，设计师在二楼墙面采用了纯正黑白格纹小牛皮毛做装饰，原始中略透露出一丝奔放；楼梯扶手以银镜饰面的柱身，让视觉效果更具有穿透力、通过反射的影像与顶上整片拉膜灯箱的呼应，从而构造出更为开敞、通透的空间，棕色木皮、石膏线型、亮色金属边框、白色磨砂漆饰面等设计元素互为映衬，相得益彰。

In order to achieve the unity of the whole and the details, the designer on the second floor wall with a pure black and white checkered calf fur decoration, slightly revealing a trace of the original in bold; stair railing with silver mirror finishes the pillar, the visual effect is more penetration through the reflection of the impact phase and pull the entire piece on top of the echo light box film, to construct a more open, transparent space, brown veneer, plaster line, light-colored metal frame, white matte paint finishes design elements against the background of each other, complement each other.

孙传进

国际资深艺术品 经纪人

当代艺术装饰类 设计师

国内资深环艺陈设艺术 设计师

中国陈设专业委员会 委员

中国室内装饰协会 会员

当代陈设艺术 设计师

近期主要荣膺

2008 年，"高尚杯"室内设计大赛二等奖；

2009年，"高尚杯"室内设计大赛一等奖；

2009年，中国建筑装饰协会设计50强大赛二等奖；

2009年，中国建筑装饰协会设计50强大赛三等奖

SuYue Restaurant
苏悦餐厅

项目面积：*1655m²*

项目地点：江苏省苏州市狮山路

设计公司：观点设计工作室

主要材料：大理石、球形花梨木饰面、玫瑰金不锈钢、实木线条、皮革等

摄影：文宗博

自古，受历史文化的影响，中国人人生的真谛在于享受淳朴悠闲的生活和社会各种关系的和睦，显现出温和、内倾的特点。公共区域设计方面，以原木色线条、大理石材、青石板 、白色鹅卵石、鸟笼及形象生动的鸟形装饰组合动态场景、宁静江南水墨景致隐于屏风种种自然元素的组合，营造自然氛围，赋予了餐厅独特魅力与亲切感。

Since ancient times, by the history and culture, the Chinese people the true meaning of life is to enjoy the simple life of leisure and social relationships of harmony, showing mild introversion characteristics. The design of public areas to wood color lines, marble material, quartzite, white pebbles, cages and bird-shaped decorative vivid combination of dynamic scenes, hidden in the quiet southern ink screen views of the various combinations of natural elements to create a natural atmosphere, giving the restaurant's unique charm and intimacy.

动与静的转换，私密的包厢内，皮革、球形花梨木饰面、现代简洁中西结合的家具形体，体现空间的华丽雅致，将传统中国元素的陈设艺术与现代材料造型相互融合，营造文化、雅致的用餐氛围。

In a private box, the leather, rosewood veneer spherical, combining Chinese and Western modern simple furniture shapes, reflecting the space gorgeous elegant, traditional Chinese elements of style furnishings and art and modern materials are blended to create a culture, an elegant dining atmosphere

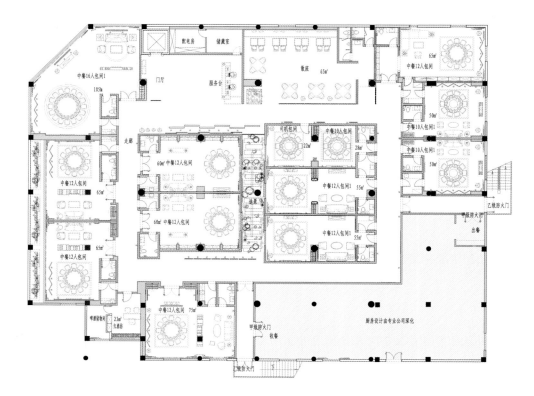

平面图/Floor plan

陈彬

武汉理工大学 副教授

中国美术家协会湖北分会 会员

CBDA中国建筑装饰协会会员及设计委员会 委员

CIID中国建筑学会室内设计分会 会员

IAI亚太建筑师与室内设计师联盟 理事

ICIAD国际室内建筑师与设计师理事会 会员

大木设计中国理事会 常务理事

后象设计师事务所 创始人及设计主持

近期主要荣膺

2011年度作品入选德国iF2012设计大奖

2011年度、2010年度作品入选英国ANDREW MARTIN国际室内设计大奖

2010年度金指环ic@ward全球室内设计大赛餐饮类金奖

2010 "中国室内设计师年度封面人物" 提名奖

2010年度IAI亚太室内设计双年大奖赛评审团特别大奖

2010年度广州国际设计周 "金堂奖" 年度十佳餐饮空间设计

2010作品荣获金外滩奖最佳餐厨空间奖

2009年度作品入选德国iF2009中国设计大奖

2009年度APIDA第十七届亚太室内设计大奖餐馆酒吧类别的最高奖项

2009 "中国室内设计年度封面人物" 提名

2009年度广州国际设计周 "金堂奖" 餐饮酒吧类中国十大设计师

2009年度中国饭店业设计装饰大赛金奖

2009年度Interior Design China "酒店餐厅类最佳设计" 奖

2009年度第七届现代装饰国际传媒奖 "年度最具潜力设计师"

2009年度IAI亚太室内设计精英邀请赛银奖

WuChang City

武昌城

项目面积: *2158m²*

项目地点: 武汉

设计公司: 大木设计中国（湖北）后象设计顾问机构 *(RHD)*

摄影: 吴辉

武昌城，建于东吴而毁于民国，存世1706年。这座近代史上的洋务重镇和辛亥名城，在漫漫的文化长河中也流淌下李白、崔颢之流的文墨风骚。

Wuchang City, established in Dongwu Dynasty and destroyed in Mingguo Period, ever survived for 1706 years. It was well-renowned as an important town both in Chinese Westernization Movement and 1911 Revolution.

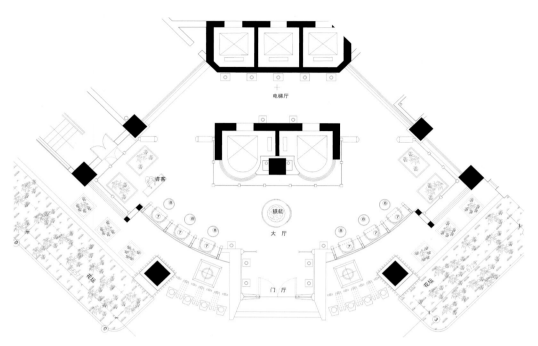

平面图/Floor plan

餐厅取名"武昌城",业主的文化诉求和高端客户定位,使设计团队对于将要创造的空间之美学追求和视觉体验有了清晰的思路和表达方向。因为对传统艺术形式的喜爱,终于决定以文人画为切入点,以当代视角剖析其审美追求。运用布局、色调、灯光、材质以及艺术陈设等方面的匠心营造,描绘出一卷"简、拙、淡、雅"的新文人气质禅意画面。

Named as Wuchang City accordingly, the restaurant particularly targets at the high-end customers with great cultural appeal, takes humanistic painting as the cut-in point and successfully offers you a literally meditative picture full of simpleness, plainness, lightness and elegance.

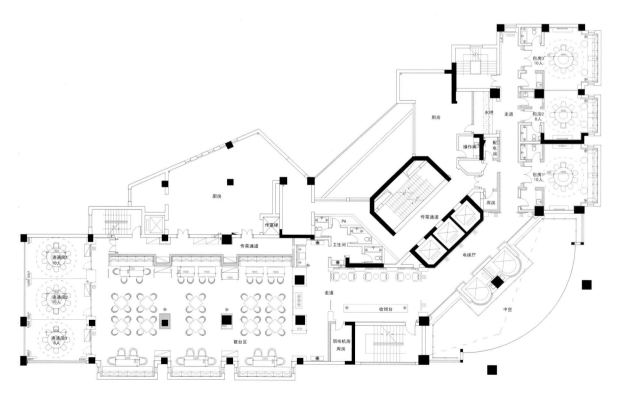

平面图/Floor plan

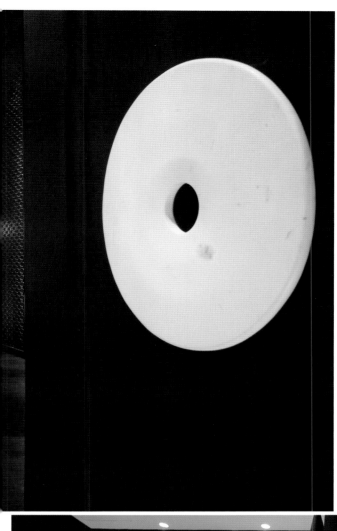

简者，造型手法，追求简约，求少求精；拙者，材质选择，饰法自然，朴实无华；淡者，色彩搭配，调性淡雅，统一和谐；雅者，饰品陈设，书卷韵味，宁静高洁。

Simpleness, seeking simple style and essence in special ways and manners. Plainness, creating natural touch and plain manifestation by material selection.Lightness, producing light, uniform and harmonious environment by way of color arrangement. Elegance, creating literal charm, quietness and nobility with decoration and furnishing.

足见，设计团队对传统文人画的内心解读在此美食空间的营造中，成功转变为物质诠释。

Right here, your inner interpretation of the traditional humanistic painting is perfectly rendered from the physical perspective in this cuisine space.

郑杨辉

福州创意未来装饰设计有限公司 设计总监
中国注册高级室内设计师、CIID设计学会 会员
中国建筑学会室内设计分会第八委员会 秘书长
中国建筑学会室内设计分会 理事

Unusual Story Restaurant

寻常故事特色餐饮

项目面积：*580m²*

项目地点：福州梅峰路

主案设计：郑杨辉

设计公司：福州创意未来装饰设计有限公司

主要材料：仿古砖、藤制品、艺术纱窗、艺术雕花

寻常故事餐饮空间，每间店铺都有一个故事，一个主题，因为生活在城市的人群，有太多的故事每天在这样的场所发生，本案的故事主题是质朴的新东方原生态空间场景再现，宁静中的奢华时尚淡淡散发，空间场景营造的主角是藤制品。材质的选择是质朴的，设计手法造型是当代的，各种藤制造型定制的灯具在空间中诉说宁静质朴，2700摄温照射角10度暖色系的照明让置身于这样的空间氛围中，所有的味觉都发挥到了极致，慢慢享用美餐的同时有了当代视觉的审美熏染。

Each restaurant shop has a story, a theme. Because people live in cities, there are too many stories happen every day in this place, the case of the story, the theme is rustic New Oriental reproduction of the original eco-space scenes, the quiet light distribution in the luxury fashion, to create a space scene the protagonist is the vine products. Material choice is simple, the style is contemporary design techniques, a variety of custom rattan lamps in the shape of the space to tell quiet rustic, 2700 intake temperature irradiation angle of 10 degrees warm colors of making clear the space so that exposure to such an atmosphere, all are to an extreme taste, delicious meal at the same time slowly with contemporary visual aesthetic he argued.

藤制品的云彩装置吊灯造型，是意念中的蓝天白云，通道的红色纱墙温暖着整个空间，只愿每个来临寻常故事的都市人群，抚摸红色纱墙柔软的质感触觉，就能记住餐厅的空间给予的无限温情，记住寻常故事不寻常的空间标识，虚隔断上红色的线绳是色彩的延伸，些许的不规则欧式镜框和中式老雕花组合在空间的墙面上，模糊了空间的界限，中式和欧式雕花图案组合的屏风，有些奢华和空间设计元素的混搭，但只要是当代和时尚的传达都是空间想要诉说的。

Rattan clouds devices pendant style, is the idea of the blue sky. The red yarn channel walls warm the entire space, touching the soft texture of red yarn touch the wall, you can remember the restaurant space given infinite tenderness. Remember the story unusual unusual space identification, virtual partitions on the color red is an extension cord, irregular frame of European and Chinese combination of old carved on the wall in the space, blurring the boundaries of space. Combination of Chinese and European-style carved pattern of the screen, to convey the emotional space.

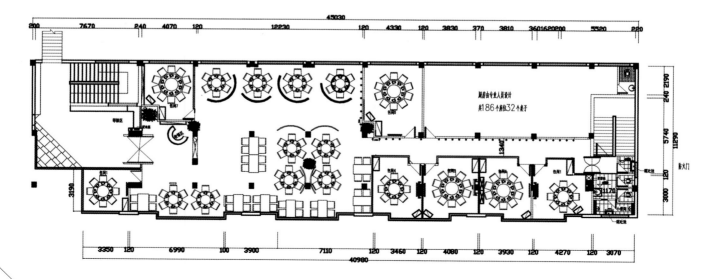

平面图/Floor plan

座位品种范围从团体到个人，有些是以低餐桌混合高餐桌，以适应不同工作人员的不同工作时间。贵宾区在尽头设置了半透明的玻璃趟门，如果有需要可以打开，形成一个更加一致的空间。

Seats range from group to individual species, some are mixed high-low table dining table, in order to adapt to different workers in different working hours. VIP area at the end of the set of translucent glass doors, you can open if there is a need to form a more consistent space.

陈彬

武汉理工大学 副教授

中国美术家协会湖北分会 会员

CBDA中国建筑装饰协会会员及设计委员会 委员

CIID中国建筑学会室内设计分会 会员

IAI亚太建筑师与室内设计师联盟 理事

ICIAD国际室内建筑师与设计师理事会 会员

大木设计中国理事会 常务理事

后象设计师事务所 创始人及设计主持

近期主要荣膺

2011年度作品入选德国iF2012设计大奖

2011年度、2010年度作品入选英国ANDREW MARTIN国际室内设计大奖

2010年度金指环ic@ward全球室内设计大赛餐饮类金奖

2010 "中国室内设计师年度封面人物" 提名奖

2010年度IAI亚太室内设计双年大奖赛评审团特别大奖

2010年度广州国际设计周 "金堂奖" 年度十佳餐饮空间设计

2010作品荣获金外滩奖最佳餐厨空间奖

2009年度作品入选德国iF2009中国设计大奖

2009年度APIDA第十七届亚太室内设计大奖餐馆酒吧类别的最高奖项

2009 "中国室内设计年度封面人物" 提名

2009年度广州国际设计周 "金堂奖" 餐饮酒吧类中国十大设计师

2009年度中国饭店业设计装饰大赛金奖

2009年度Interior Design China "酒店餐厅类最佳设计" 奖

2009年度第七届现代装饰国际传媒奖 "年度最具潜力设计师"

2009年度IAI亚太室内设计精英邀请赛银奖

YinLu Restaurant

隐庐

项目面积：1000m²

项目地点：武汉

设计公司：大木设计中国（湖北）后象设计

顾问机构 (RHD)

摄影：吴辉

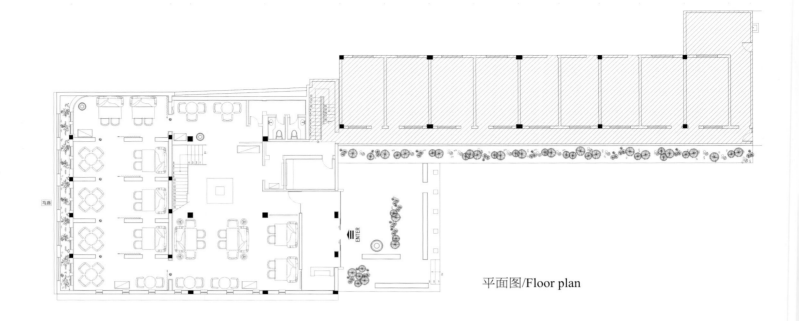

平面图/Floor plan

选址在略微喧闹的街道与业主追求安静平和的用餐气氛以及偏高端的市场定位似乎有着不可调和的矛盾。设计工作的重点是找到一种方法使这个餐厅空间真正的能够"闹中取静"。设计师及团队决定以建筑外观形态、空间动线规划和界面的视觉暗示这些方面为切入点来完成这项工作。

Seeking a way to make this restaurant a real space for stealing quite and peaceful mind from the hustling city life,our design takes the architectural exterior form, spatial axial line planning and interface visual hint as the key points.

大面积平整的水刷石墙面和特意缩小加长的窗形，以及模仿自然界晨雾的玻璃丛竹夹层，使建筑在不牺牲采光的前提下，尽可能的隔绝外界的干扰。而将餐厅入口背离街面，开在后院并设置一个竹林小院的动线规划更是设计师"取静"的神来之笔。

Exteriorly, large-area water-brushed stone walls are selected to make the building facades simplified and the restaurant entry is redesigned to be located in the backstreet. Interiorly, serene terrazzo floor, amicable wooden wall and nostalgically decadent handmade tiles are selected to create a serene touch.

室内空间中深色水磨石地面和实木墙板及家俬的运用，使整个餐厅充满湖水的宁静和草木微语，而一种冷色调手工砖的使用，又给这个空间注入一丝怀旧的颓废气息。完成后的餐厅，自然、平淡、静谧，渗透着一种东方内敛的精致。所谓"大隐住朝市，小隐入丘樊"，餐厅取名"隐庐"。

Upon completion, the restaurant, permeated with a kind of oriental introversive refinement, fully displays thorough naturalness, plainness and serenity. As a poem goes: The most capable people stay in worldly life while those seeking for peaceful mind stay in Hidden Cottage which is remote and quiet. The restaurant is therefore named "Hidden Cottage".

该案在选材上全场采用LED节能控制系统，墙面大量采用环保涂料，色彩协调，局部采用不锈钢及铜片，营造一个舒适的用餐环境。

Seats range from group to individual species, some are mixed high-low table dining table, in order to adapt to different workers in different working hours. VIP area at the end of the set of translucent glass doors, you can open if there is a need to form a more consistent space.

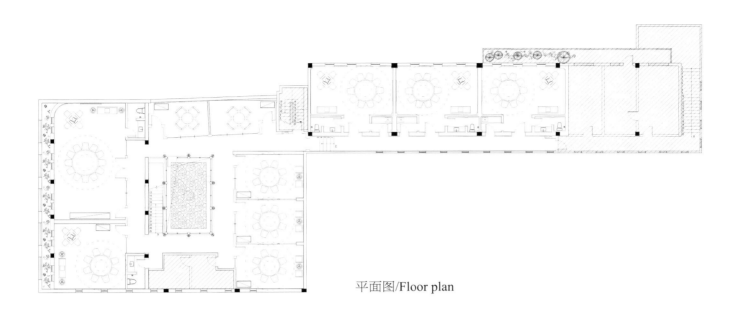

平面图/Floor plan

林金华

2002年毕业于福建师大美术系

2002年到2009年任合诚环境艺术设计有限公司 主任设计师

2010年参与创办楼语设计顾问工作室

Fisherman's Wharf Hotpot

渔人码头时尚火锅店

项目名称：渔人码头时尚鱼火锅

面积：**680m²**

设计：林金华

主要材料：仿锈砖、墙纸、造型藤灯、大理石

施工单位：朗雅别墅施工

摄影：施凯

进入渔人码头的时尚火锅店中，藤灯与防锈砖、大理石与墙纸交糅在一起，既矛盾又和谐，有一种苍劲的美，又似乎若隐若现地展示出一缕淡定的情愫，给人予视觉上的冲击力。设计师在进门的大厅处放置了一个红色的、憨态可掬的人物雕塑，它既起到了迎客的效果，也与前台背景墙上的书法装饰形成刚柔对比，不着痕迹而又颇具用心地将传统气息沁入其间，即使是浮躁的心绪也会在不知不觉中安静了下来。

Into the restaurant, rattan lamps and rust-proof brick, marble and wallpaper pay melted together, contradiction and harmony, there is a vigorous beauty, it seems to show a ray of looming calm affair, giving to the visual impact . Designers in the door of the hall, placed a red, naive sculpture, both played a welcoming effect, but also the background with the front wall of the formation of rigid decorative calligraphy contrast, implied but quite hard to the traditional atmosphere Qinru the meantime, even the fickle mood will unknowingly quiet.

在用餐区域中，既有卡座，也有圆弧形的小包间。带来了奇妙而趣味十足的感官享受。在楼梯处，相似造型的蓝色装置陈设则给人别样的视觉感染力。在整体的暗环境中，这些光影的变化使得空间里的气息有了一丝流动的意味，仿若重复滚动播放的胶片上，闪烁着海市蜃楼的身影，给人如此微妙又似曾相识的体验，激发出一种愉悦的想象。

Restaurants in both decks, but also between the arc-shaped packet. Brought a wonderful and fun and full of sensuality. In stairs, similar to the shape of the blue device gives a different kind of display is the visual appeal. In the overall dark environment, these changes in light and shadow makes the space atmosphere with a trace of the mean flow, are like the film repeatedly scrolling, flashing a mirage of the figure, giving such a delicate and familiar experience, inspire a kind of pleasure of the imagination.

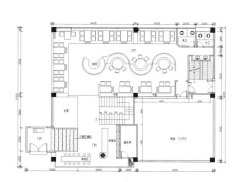

一楼平面图/1st Floor plan

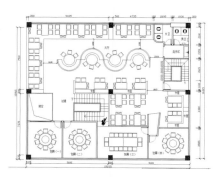

二楼平面图/2nd Floor plan

独立包厢中，设计师或以蓝色的背景装饰营造水的气息，或以黄色的挂画带出温暖的质感。这些丰富的元素融入周遭环境，令静者之间的丰富关系。

Separate box, the designer or blue background decorated to create the atmosphere of water, or paintings with a warm yellow texture. These rich elements into the surrounding environment, so that the static relationship between the rich.

孙华锋

高级室内建筑师

中国建筑学会室内设计分会 常务理事

中国建筑学会室内设计分会第十五专业委员会
主任

河南鼎合建筑装饰设计工程有限公司 总经理

河南省中青年专家 组专家

《INTERIOR DESIGN CHINA》杂志郑州联络处
主任

刘世尧

高级室内建筑师

中国建筑学会室内设计分会第十五专业委员会
副主任

美国IFDA国际室内装饰设计协会河南办事处
主任

河南省建筑装饰协会家装委员会 副秘书长

河南鼎合建筑装饰设计工程有限公司 执行董事

Authentic Japanese Kaiseki

四季怀石料理

坐落地点：河南郑东新区

建筑面积：*1500 m²*

设计师：刘世尧、孙华锋

参与设计：李珂、张利娟、孙健

主要材料：黑白玉大理石、红影木、毛石、黑色铁
艺，草编壁纸、实木花格

本案餐厅主要为品味高雅的顾客提供正宗的日式料理。本案从现代的美学角度让日式文化获得了新的生命，在整个空间中彻底琢磨锤炼每一个细节，使人强烈感受到材料的质感与力道，让顾客在现代的用餐环境中体味传统的日式料理。

The case mainly for taste and elegant restaurant customers authentic Japanese cuisine. The aesthetics of the case from the perspective of modern Japanese culture to get a new life in the whole space, every detail thoroughly refined temper, people strongly feel the texture and strength of materials for customers in the modern dining environment, appreciate traditional Japanese cuisine.

整个空间格局清晰，从接待区到寿司台区再到铁板烧及榻榻米包间，各个空间都有其独特的设计和别致的艺术品装置。从一层门厅开始，就对日式元素进行了提炼。上到二层，穿过接待区就是主用餐区。3.4米的挑高和临窗的位置使这里拥有极佳的视野，座位间的黑色铁艺隔断既让客人感到相对的私密性，又让这一区域多了一份安静。粗犷，自然的毛石墙面成为视觉焦点，订制的日本纸灯和艺术装置，也将整个餐厅的品味体现出来。铁板烧上方飞舞的樱花图案让整个区域凭空多了一份浪漫，大理石台面上雅致的黑白色水纹，与走道波浪状的黑色隔断相呼应，让空间格调更加统一。

The spatial pattern of clear, from the reception area to the station area to the teppanyaki and sushi tatami rooms, each room has its own unique design and unique art installations. Floor lobby from the beginning, on the Japanese elements were extracted. On the second floor, through the reception area is the main dining area. 3.4 m high ceiling and by the window here has an excellent position to make the vision of the black wrought iron seating guests feel cut off both the relative privacy, letting more of a quiet region. Rough, natural rubble walls become the visual focus of the Japanese paper lamps and custom art installations, the taste will be reflected throughout the restaurant. Sakura Teppanyaki fluttering above the pattern out of thin air so that the entire region more than a romantic, elegant black and white marble countertop Watermark, with wavy black aisle echoes cut off, so that space is more uniform style.

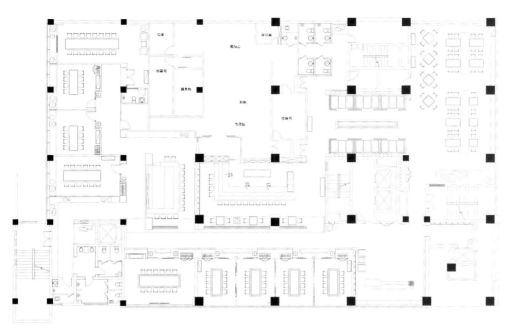

平面图/Floor plan

凌子达

KLID达观国际 设计总监

台湾逢甲大学建筑系毕业

法国**CNAM** 国立工艺学院 设计管理硕士

近期主要作品

万科深圳——松山湖样板房设计；

万科东莞——万创样板房研发设计；

万科珠海——售楼处与样板房设计；

金地上海——格林世界最后期别墅样板房设计；

盛清上海——北竿山别墅样板房设计；

建发长沙——汇金别墅样板放设计；

泛华厦门——售楼处建筑与室内设计；

永年厦门——售楼处建筑与室内设计；

万通南通——售楼处建筑与室内设计与样板房设计；

博泰南昌——办公楼设计，等等。

Haneda Japanese Restaurant

羽田日本料理餐厅

坐落地点：福州

建筑面积：*600* m²

设计公司：*KLID*达观国际

设计师：凌子达

主要材料：钢材、耶路撒冷灰石材、水泥、集成材

摄影：周耀东

一个完美的美食体验，首先重要的是精致的美食，同时还要能够配合心灵层次的空间感受。视觉与味觉的完美结合，就是最佳的体验。

在整体设计上追求着一种宁静的感觉，运用混凝土肌理的表现，期望塑造出一种沉稳内敛的感觉。入口处放上了一块很自然的原石，由原石的缝隙中静静的流水，代表着静中有动的生命力。许多铁艺的设计是使空间有更丰富的层次。

A perfect dining experience, the first important thing is fine dining, but also to meet the spiritual level of feeling of space. The perfect combination of visual and taste, is the best experience.

 In the pursuit of the overall design feels a kind of quiet, the use of concrete texture performance expectations create a kind of steady, introverted feeling. Placed on a natural entrance to the original stone, the gap in the original stone quiet water, representing the static in action vitality. Many iron is designed to make space for more extensive level.

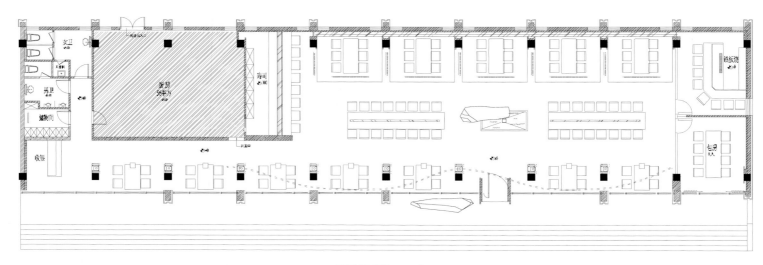

平面图/Floor plan

陈轩明

香港大学建筑系毕业

香港中文大学可持续建筑设计 环境硕士

澳洲墨尔本理工大学 艺术硕士

近期主要荣膺

筑巢奖2010中国国际空间环境艺术设计大赛三等奖 —— 北京首都时代广场地铁通道（2010）；香港嘉禾青衣电影城（2010）

筑巢奖2010中国国际空间环境艺术设计大赛优秀奖 —— 香港嘉禾荃新电影城（2010）

"欧普·光·空间杯"办公空间照明应用设计大赛中国top10办公空间照明应用设计年度人物奖（2010）

"欧普·光·空间杯"办公空间照明应用设计大赛"最佳创意空间奖" —— 美丽华酒店办公室（2010）

亚太室内设计双年大奖赛入围 —— 香港嘉禾青衣电影城（2010）

最成功设计奖 —— 香港嘉禾荃新电影城（2010）

iF 中国区设计奖 —— 香港嘉禾青衣电影城（2010）

星火国际设计奖 —— 嘉禾青衣电影院，新城国际样板间，维健牙医诊所（2009）

最成功设计奖 —— 嘉禾青衣电影院（2009）

First Round- Hong Kong Contemporary Art Biennale（2009）

亚太室内设计大奖十名入围商业类 —— 香港嘉禾青衣电影城（2009）

Miramar Restaurant
美丽华酒店餐厅

项目面积：**355m²**

项目地点：香港九龙尖沙咀弥敦道

设计公司：*DPWT Design Ltd*

主要材料：清镜、不锈钢、拼花木地板、暖色调子喷漆墙身、塑料薄膜照明

摄影：陈志威

美丽华餐厅以暖色调子为核心：红橙黄三色渗透整个空间。这种鲜明的色彩结合空间，同时创造一个高能量的感觉。为了配合良好的整体气氛，整个餐厅采用白色家具，营造一个别致而难忘的印象。鉴于该餐厅的位置在地下层，没有任何窗户和采光。因此，在室内引进一些镜子来提高空间的质量。镜子通常是安装在远处的墙上，通过反射，空间可以增加一倍的视觉假象。DPWT 为墙壁和隔板设计了一系列"跳舞"图形，从这家餐厅的主题基础，证明了它的适应性，通过重要的方案以不同多变的面貌重新表现。作为一个视觉趣味，创造一个引人注目的12米长的视觉艺术壁板，并在离地面1米左右的空间展出。

Miramar restaurant to warm sub-core: red orange tricolor penetrate the entire space. This vibrant color combination of space, while creating a high-energy feeling. In line with good overall atmosphere, the restaurant with white furniture, creating a unique and unforgettable impression. Location of the restaurant in the basement, no windows and lighting. Therefore, the introduction of a number of mirrors in the room to improve the quality of space. DPWT walls and partitions for the design of a series of "dance" graphics, the theme from this restaurant based on proven its adaptability to different important programs by changing the face of re-performance. As a visual interest, creating a striking 12-meter-long wall of the visual arts, and from about 1 m from the ground space for display.

照明设计分为两个区域。首先，在酒吧区创作白色流线形状天花面板配合球状吊灯衬托地板的分隔线，制造浮动和轻松的环境。另一方面，进餐区的天花采用线状和规则形状的照明嵌板以非传统的形式排列组成，给观众提供了一个更加具活力和生气的组合。

Lighting design is divided into two regions. First, create a white area in the bar-shaped ceiling panels streamline chandelier ball off the floor with the dividing line, making floating and relaxed environment. On the other hand, dining area and ceiling rules using linear shape of the lighting panels composed of non-traditional arrangement, provided viewers with a more vibrant and a combination of anger.

SCALE 1:200

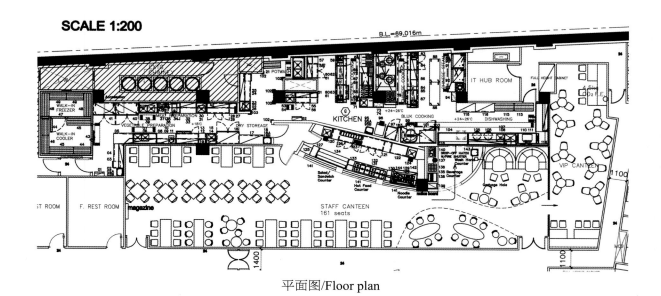

平面图/Floor plan

本案从和业主探讨开始，就定位为时尚餐厅，菜品以创新中西餐为亮点，设计采用混搭的风格元素，色彩大胆、抢眼，带来全新的餐饮体验。和同类餐饮相比，环境风格上是中西文化的一个结合，是一次古典和现代，西方与东方文化的创新。

餐厅在空间布局上考虑到不同消费人群的需求，具有开放空间和私密空间的互补，同时还开辟一个酒吧区，可以在用餐后小酌。在选材上，全场采用**LED**节能控制系统，墙面大量采用环保涂料，色彩协调，局部采用不锈钢及铜片，营造一个舒适的用餐环境。

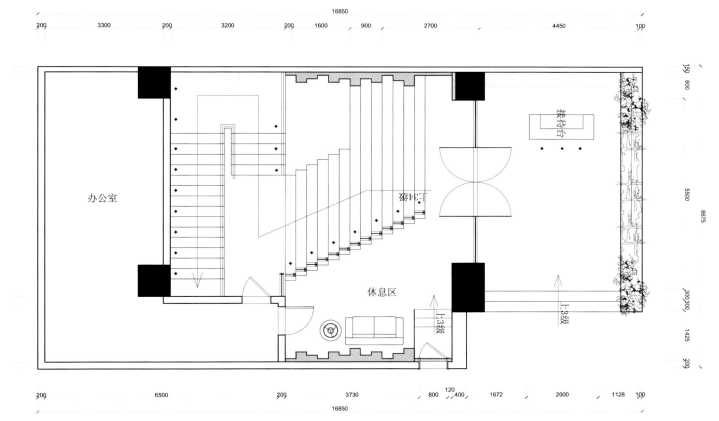

一层平面图/1st Floor plan

办公室

休息区

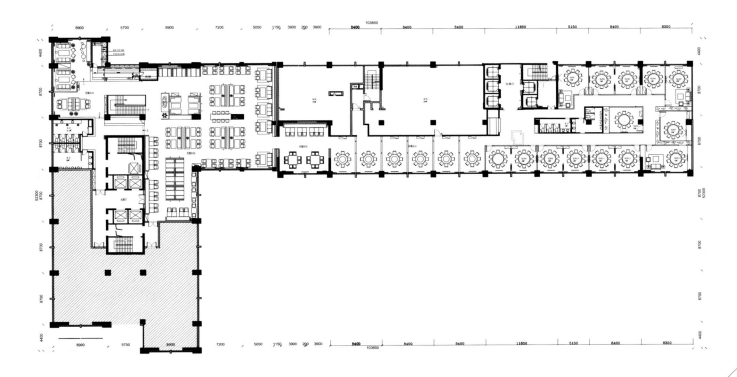

二层平面图/2nd Floor plan

And the owners of the case from the beginning, it positioned as a trendy restaurant, dishes for the bright spot in the Western innovation, designed with mix and match elements of style, color, bold, eye-catching, bringing a new dining experience. Case and compared to similar restaurants, environmental style is a combination of Eastern and Western cultures, is a classical and modern, Western and Eastern culture of innovation.

Taking into account the spatial distribution of different consumer groups, with open space and private space, complementary, and also open up a bar area, you can drink after a meal. The case in the selection of the audience on the use of LED energy-saving control systems, extensive use of environmentally friendly wall paint, color coordination, partial use of stainless steel and copper, creating a comfortable dining environment.

张健

毕业于杭州某美术类学校

1999年起，从事室内设计工作

2007年底，创立杭州观堂设计

近期代表作品

蜜桃咖啡

青桃餐厅

7080

迪赛尼斯女装旗舰店

Yongle Fang Restaurant

永乐坊餐厅

项目名称：永乐坊餐厅

坐落地点：杭州市下城区河东路

设计公司：杭州观堂设计

参与设计：陈蕾、元立俊

面积：**700m²**

主要材料：回收木板等

摄影：王飞

业主想打造一个主题明确、可以让60、70、80年代人产生共鸣的餐厅，因此设计过程中挖掘了很多那个年代特有的元素。

Owners want to create a subject-specific, so that people can resonate with the restaurant, so the design process that's digging a lot of unique elements.

大厅区块，餐桌餐椅都是木质，或采用原木色，或刷成白漆，仿佛年少时的课桌。包厢区块，用曾经的课本命名，语文，数学，音乐，每个包厢因为主题的不同，内部陈设也完全不同。推开一扇门，或许铺满了报纸，或许挂满了吉他、CD，或许满是地球仪、望远镜。

雷锋、少先队、牛皮筋、造句、毛主席画像、双卡录音机，太多的内容，都可以在这家主题餐厅得以显现，借以回忆我们无处安放的青葱岁月！

Hall block dining table is wood, or the use of wood color or painted white paint, as if childhood desk. Box block, with the textbooks we have named, language, mathematics, music, because the theme of each box is different from the internal into the set is completely different. Open a door, may be covered with newspapers, perhaps covered with a guitar, CD, perhaps full of globes, telescopes.

Lei Feng, the Young Pioneers, leather bars, sentences, pictures of Chairman Mao, dual card recorder, a lot of content, can be seen in this theme restaurant, cause us endless memories.

Dariel & Arfeuillere

Thomas Dariel 和 Benoit Arfeuillere是总部位于上海的创意机构 Lime 388 的创立者

2010年，获得"年度最具潜力设计师奖"

近期主要作品

The Funky Chicken

Yucca（金堂奖 2011 China-Designer 中国室内设计年度评选 年度十佳优秀餐饮空间设计）

12 Chairs

The Fat Olive

The Purple Onion

Kartel酒吧

The Funky Chicken
The Funky Chicken餐厅

项目名称：*The Funky Chicken*

坐落地点：*上海思南公馆26F*

设计公司：*Dariel & Arfeuillere*

——*A Lime 388 Company*

设计师：*Thomas DARIEL & Benoit ARFEUILLERE*

主要材料：乳胶漆，人造石，马赛克

The Funky Chicken带来了时髦的烤鸡快餐。由卡通和搞笑智慧激发灵感，Dariel & Arfeuillere 运用了大胆的色彩及出乎意料的建筑元素打造了一个有趣，刺激的零售环境。

The Funky Chicken brought a trendy fast food chicken. By the cartoons and funny wisdom inspired, Dariel & Arfeuillere use of bold colors and unexpected architectural elements to create a fun, exciting retail environment.

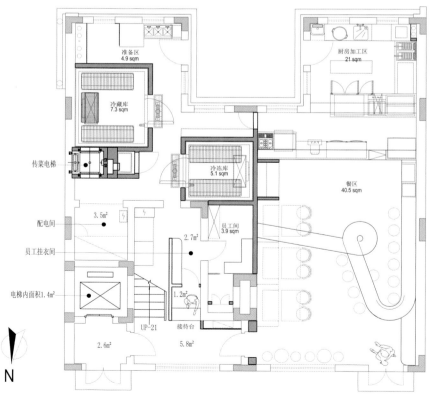

平面图/Floor plan

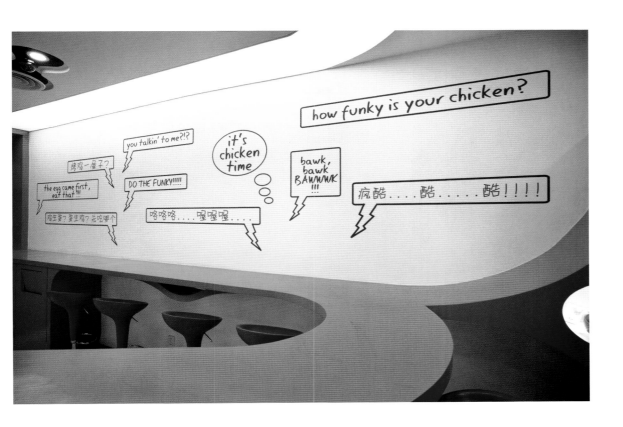

墙上的涂鸦写着："乱叫个啥"另一面墙上则印着两只鸡不可思议的对话。正中的红色坐位台面围绕餐厅旋转而上，受到了传统西方农庄及飞机的启发。

Graffiti reads: "barking Gesha" the other side of the wall is printed with two chickens incredible dialogue. Around the middle of the red seat dining table rotation, and by the traditional Western farm and aircraft inspiration.

仲松

1999年 毕业于中央美术学院雕塑系

1994年 毕业于中央美术学院附中

近期主要作品

2010年，江西彭友善美术馆建筑、室内、景观设计

2010年，武夷山钰荣大酒店建筑、室内、景观设计

2009年，上海龙美术馆总体设计

2009年，杭州阿曼酒店总体设计

2008年，国酒茅台会所室内及景观设计

2008年，上海五角场800号美术馆总体设计

2007年，北京SONG CLUB室内设计

2007年，北京宋庄SOKA私立美术馆设计

Lemarais Restaurant

蓝玛赫西餐厅旗舰店

建筑面积：*1178m²*

项目地点：深圳市华侨城欢乐海岸东区8#建筑商铺

设计单位：北京仲松建筑景观设计顾问有限公司

主持设计：仲松

参与设计：吴庆东、陈国良

摄　影：井旭峰

项目前期，设计团队充分参与了Lemarais Restaurant品牌的再定位，以及服务系统的完善等全程的策划工作，改变原有餐饮品牌差异性不足等劣势，同时通过本次对Restaurant的设计确立全新而鲜明的品牌形象，以适应其未来全国连锁的品牌发展策略。

本案在设计上将餐厅运用单一构造的木质隔断巧妙划分出疏密有秩的若干功能区域，形成自由丰富且连续性的空间秩序。

Early in the project, the design team to participate fully in the Lemarais Restaurant brand re-positioning, as well as the improvement of other full service system planning effort to change the existing lack of differences in food and beverage brands disadvantage, through the design of this set of new and Restaurant distinctive brand image, to suit their national chain brand in the future development strategy.

Restaurant use in the design of the case will be constructed of a single partition skillfully carved out of wood density and orderly several functional areas, rich and continuous formation of free space in order.

设计的核心在于选择与中国文化密切相关的经典物料——"木"作为其基本元素,然后用惊人的数量重复运用并营造一种令人震撼的视觉景观——"竹林"。

Selection and design of the core is closely related to Chinese culture classic material - the 'wood' as its basic elements, then repeat with the amazing amount of use and create a stunning visual landscape - "Bamboo."

以唤起人们对已形成的形象识别
的熟悉意向的记忆，使人的行为
融入这幻化出的"纯净"中，并获
得强烈的真实感。

Has been formed to raise aw-
areness of the familiar image
of the intention recognition
memory, people's behavior
into the metamorphosis of this
"pure" in and get a strong sense
of reality.

XiangLai Concept Restaurant

香籁概念餐厅

项目地点：武汉市菱角湖万达广场

设计单位：武汉品筑凌川设计顾问有限公司

主持设计：凌川

建筑面积：530m²

主要材料：软膜、银白龙大理、石、灰镜、镜钢

摄影师：吴辉

香籁概念餐厅是一家概念中餐美食料理餐厅，位于万达商圈。走进这个空间，行云流水般的线条强烈的解构主义美学，让人几乎忘却了这原本是个餐厅。空间面积并不大，却以流线型的隔断环围出一个个私密的空间，空间的特色造型极具视觉冲击力，让人几乎忽略了实际空间面积的大小。

The concept of Chinese food restaurant is a gourmet eatery. Into this space, the strong lines to fill the gap deconstruction aesthetics, people almost forget this was originally a restaurant. Space area is not large, but the streamlined cut off a ring around a private space, the space features shape great visual impact, people almost ignore the actual space the size of the area.

尽管极富挑战，香籁概念餐厅最终还是以完美无瑕的品质赢得了业主及其来往客户的厚爱。本案设计极具纪念意义，在餐饮设计业竞争激烈的中国，武汉设计师锐意进取、突破陈规，本案营造的时尚优雅的气氛，堪称典范。

Commemorative significance of the case design, restaurant design industry is highly competitive in China, Wuhan designers forge ahead, breaking the stereotypes, he gave the case to create a stylish and elegant atmosphere, exemplary.

张健

毕业于杭州某美术类学校

1999年起，从事室内设计工作

2007年底，创立杭州观堂设计

近期代表作品

蜜桃咖啡

青桃餐厅

7080

迪赛尼斯女装旗舰店

Metoo Café

METOO Café杭州滨江店

面积：600m²

坐落地点：杭州市滨江区滨盛路4309号

设计公司：杭州观堂设计

设计：张健

主要材料：各类回收材料

摄影：王飞

由于建筑面积较大，考虑成本，设计师再一次大力运用了环保的手法，二楼地面、吧台均采用回收木板，再加工后投入使用。抬头望天，会发现顶面上"长"上了各种老式的皮箱，这些皮箱也是设计师从各处精心搜罗过来，一是作为别出心裁的装饰，时空颠倒，老式的旧皮箱跑到了吊顶上，给客人带来感官上的强烈震撼；二是解决了裸顶布线难看及灯光灯源的问题，从而保持了顶部的整体干净；三是以极低的成本完成了顶部装饰，并贯彻了环保的理念。

Construction area, consider the cost, the designer once again vigorously the use of environmentally friendly practices, the second floor of the ground, bar the use of recycled wood, and then processed and put into use.Looked up to heaven, will find that the top surface of a variety of old-fashioned suitcase, the suitcase is also designers from a variety of carefully collecting over a result, as the decoration of the ingenuity, time and space upside down, old-fashioned old suitcase went on the ceiling.bring a strong shock on the sensory to the guests; two to solve the problem of lighting light source, thereby maintaining the top of the overall clean; three to a very low cost to complete the top of the decorative and implement the concept of environmental protection.

METOO Café滨江店里的所有内容均采用回收材料，一楼至二楼的走廊通道上，全是老门板，各种花纹、各种尺寸、各种颜色，设计师充分运用深厚的平面设计功底，将他们拼成整面的墙与顶，天与墙连为一体，给客人强烈的体验；二楼墙体上，采用了回收红酒瓶，经过特殊的处理，两面夹上玻璃，便是一个漂亮却又简单的半透明隔断；餐厅内的桌面、柜体都采用回收木材制成，在柜子的把手上做了一些细节，用乒乓球拍、刀、叉做把手，让人在不经意间有所感动。

METOO Café where recycled materials are used, the first floor to the second floor of the corridors, and all the old door, a variety of patterns, sizes, colors, designers make full use of strong graphic design skills, makes up the entire surface of their walls and top, the sky and the wall as a single entity, to the intense experience of the guests; on the second floor wall, with a recovery of the red bottle, through a special deal, both sides of the clip on the glass, there is a pretty simple translucent partition; restaurant the desktop, cabinets are made of recycled wood, the handle on the cabinet to do some detail, with table tennis bats, knives, forks do hand, people in casual feel anything.

软装上，也大量采取了回收材料；比如进门处的老款电视机，全部刷成白色堆放成圈；吧台上方的老式玻璃灯罩，各种款式，各种颜色，在设计师手中变得格外迷人；我们年幼时所接触的印花铁皮饼干盒，若干个排列在柜子上，透露出幸福与岁月的痕迹。

Software installed, it took a lot of recycled materials; such as the entrance of the old television sets, all piled into a circle painted white; bar at the top of the old-fashioned glass lamp shade, a variety of styles, colors, in the hands of the designer has become particularly charming; when we come into contact with young stamp tin biscuit boxes, arranged in a number of cabinet, revealing traces of the years.

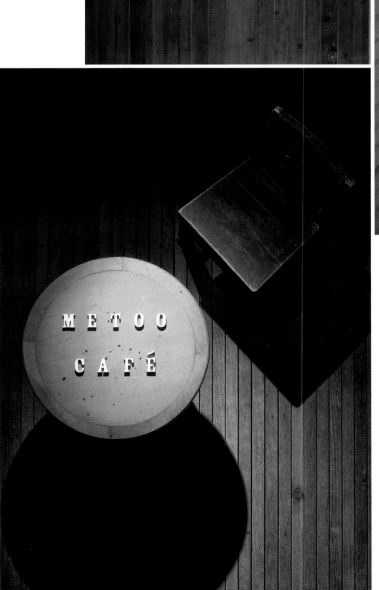

二楼中间,设计师特意搭建了一个木头小屋,
半敞开,小屋窗外有一圈花坛,可以摆放小小
的植物,给在内就餐的客人一种凭栏而眺的
感受,又保留了一定的私密性;同时,在必要
的场合,这座小屋可以很方便的拆除与搭建,
方便大型活动时的场地所需。

The middle of the second floor, the
designers deliberately built a wooden hut,
half open, hut circle flower bed outside
the window, you can put a small plant, to
lean on a railing, including dining guests
a feeling of the balcony, but also to retain
a certain amount of privacy nature; the
same time, where necessary, this house
can easily remove and set up, easy to
need a large venue events.

黄振耀

意大利米兰理工大学室内设计管理学硕士
清华大学酒店设计高级研究生班毕业
IAI亚太建筑师与室内设计师协会福建分会副理事
IDA国际设计师协会分会副会长
中国建筑协会室内设计分会资深会员

主要作品

IDA2008国际建筑景观室内设计大奖赛样板房类
特等奖

IDA2008国际建筑景观室内设计大奖赛餐饮类金
奖

2009国际建筑与室内设计节金外滩奖提名奖

2009国际建筑与室内设计节金外滩奖4个入围设
计奖

17届香港亚太室内设计大赛酒店空间前十名

2009中国室内空间环境艺术设计大赛酒店类三等
奖

2010年中国（上海）国际建筑及室内设计节金外
滩"最佳概念设计奖"

2010 "INTERIOR DESIGN China 酒店设计
奖——最佳酒店设计奖"

作品录入《2009优秀室内设计师》《中国创意
界》《17届香港亚太室内大奖入围及获奖作品
集》等相关作品集

Pinxi Jingshe Restaurant

品汐精舍

坐落地点：厦门软件园二期

设计单位：东峻国际设计顾问

设　计：黄振耀

建筑面积：**450 m²**

主要材料：原木、硅酸钙板

摄　影：黄振耀

本案遵循"敬重自然""因地制宜"的低碳原则做设计。由于依托在酒店的后花园，周边环境景色宜人、绿意葱葱。如何使景观与室内空间水乳交融，形成"我中有你""你中有我"的"忘我"意境将是本案设计的关键。我们尝试用"天然，质朴"的材料，唤醒自然界中的"物灵"，让一切都充满生机。

Works follow the "respect for nature," "specific" carbon principles do the design. Because relying on the hotel's back garden, scenic surroundings, lush green. How to landscape and interior space design harmony is the key to the case. We try to use "natural, rustic" materials, wake up in nature "material spirit" so that everything is full of life.

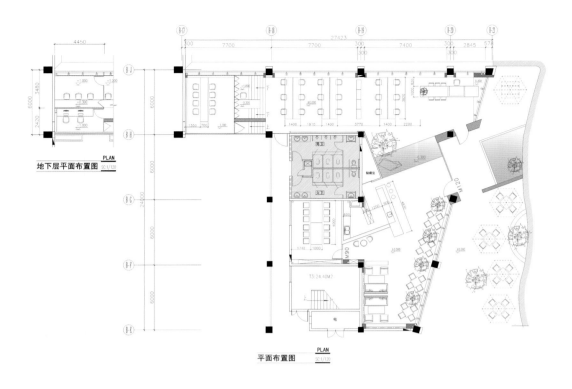

地下层平面布置图 SC:1/120

PLAN

平面布置图 SC:1/120

PLAN

空间的一半是公司办公空间，一半是半营业状态的实验餐厅。两者如何相互应用、互相拓展、交替使用，是该设计的难点。功能及动线设计上都应同时考虑两者结合应用上的灵动性与方便性。整个案子主要围绕空间结构美设计，没有多余装饰性设计元素。

Half the space is the company's office space, half of the half-open state of the experiment restaurant. Between the two applications can be expanded with each other, alternately, the difficulty of the design. Moving line design features and at the same time should consider a combination of both applications on the Smart and convenience. Spatial structure of the whole case centers around the design, no superfluous decorative design elements.

实现一颗木头摆平"一切"的初步理念。所有的家具、墙板造型由一棵长度11.7m，直径0.9 m的原木在没有一点浪费的前提下，合理切割，用可持续使用的安装方法拼装，实现真正意义上的低碳、环保及可持续应用的原则。

To achieve a wood to settle "all" of the initial concept. All the furniture, wall shape consists of a timber processing, making sustainable use of the installation methods assembled to achieve a real sense of low-carbon, environmentally friendly and sustainable application of the principle.

木头处理细节也根据古法进行制作，用最简单最质朴的方式加工，表面手工木蜡油让木头的"物灵"充分被唤醒，真正做到环境才是空间主人的"无我"意境。

Wood processing details are based on the traditional method for making, using the simplest and most simple way of processing, wood wax to the surface of the wood by hand, "things spiritual" wake up fully, truly the master of the space environment is "no" mood.

张猛

武汉U80设计事务所创始人

Seasons Restaurant

北京凤凰城四季恋餐厅

坐落地点：北京三元桥华润凤凰城

设计单位：武汉U80空间

设计师：张猛、袁宏超

建筑面积：2200 m²

主要材料：白色大理石、玻璃、墙纸、不锈钢、肌理漆

摄影师：吴辉

本项目位于北京东三环三元桥附近的华润凤凰城，该楼自身造型很有特点。所以，当我们去现场勘察时，首先就确定要利用建筑本身的特点来进行布局。

原建筑的层高很高，为了实现客位数量的保证和空间层次的变化，我们对内部空间进行了改造和加建，让室内的空间更大气和丰富。

The project is located near Beijing Sanyuan Bridge Resources Phoenix, the building itself is very characteristic shape. So, when we went to the scene investigation, first to determine the characteristics of the building itself to use for layout.

The original high-storey building, in order to guarantee the number of class and space-level changes, so we transform the internal space of the processing building, so more interior space and rich atmosphere.

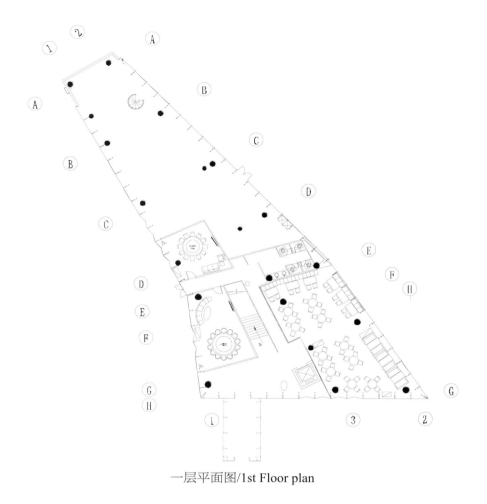

一层平面图/1st Floor plan

室内的灯光采用了大量的间接照明，烘托温馨而浪漫的气氛。用以直接照明的射灯则安排在菜品的正上方，让菜品在客人面前显得更加秀色可餐。

Interior lighting with a lot of indirect lighting, heighten the atmosphere warm and romantic. Position for direct lighting spotlights dishes are arranged just above, let dishes before the guests even more of eye candy.

餐厅的主色调用大面积灰、
白铺设，点缀不同的金色，
如玫瑰金、香槟金、砂金等，
来体现高贵而雅致的格调。

The main dining area with
large gray color, white roll,
together with the different
gold embellishment, such
as rose gold, champagne
gold, gold dust, etc., to
reflect the noble and
elegant style.

二层平面图/2nd Floor plan

Dariel & Arfeuillere

Thomas Dariel 和 Benoit Arfeuillere是总部位于上海的创意机构 Lime 388 的创立者

2010年，获得"年度最具潜力设计师奖"

近期主要作品

The Funky Chicken

Yucca（金堂奖 2011 China-Designer 中国室内设计年度评选 年度十佳优秀餐饮空间设计）

12 Chairs

The Fat Olive

The Purple Onion

Kartel酒吧

The Fat Olive
The Fat Olive餐厅

项目名称：*The Fat Olive*

坐落地点：上海思南公馆26F

设计公司：*Dariel & Arfeuillere*

——*A Lime 388 Company*

设计师：*Thomas DARIEL & Benoit ARFEUILLERE*

主要材料：防腐地板、灰镜、大理石

摄影师：*Derryck Menere*

The Fat Olive 融合了希腊家庭风格。由欧洲乡村市集激发灵感， 设计师将其打造成放松舒适的场所。"我们想打造家的感觉，因而我们选择了有质感和历史感的材料和细节"。

设计师通过选择设计元素达成了这种效果。木制地板强调了欧洲居家感觉，从欧洲乡村市集中得到灵感启发的酒柜墙，长毛绒淡紫中产阶级情调沙发散发出中庸的生活气息。

Case of fusion of the Greek family style restaurant. Inspired by the European village marketplace, designers will fight the relaxed and comfortable place. "We want to create a feeling of home and therefore we have chosen the material texture and a sense of history and details."

Designers achieved by selecting the design elements of this effect. Wooden floor feeling stressed at home in Europe, European country to be inspired by the city concentrated wine cooler walls, plush purple sofa exudes a moderate middle-class flavor the breath of life.

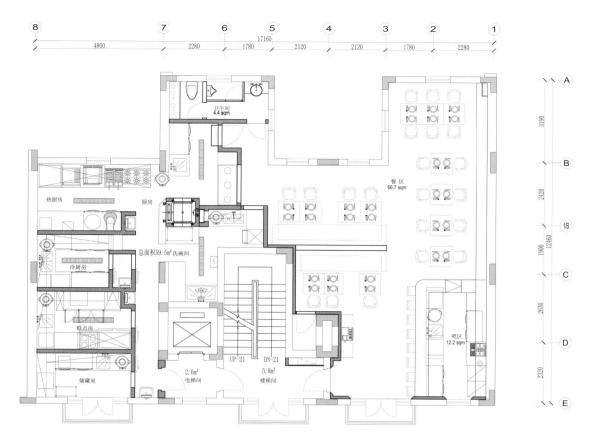

平面图/Floor plan

Dariel & Arfeuillere

Thomas Dariel 和 Benoit Arfeuillere是总部位于上海的创意机构 Lime 388 的创立者

2010年，获得"年度最具潜力设计师奖"

<u>近期主要作品</u>

The Funky Chicken

Yucca（金堂奖 2011 China-Designer 中国室内设计年度评选 年度十佳优秀餐饮空间设计）

12 Chairs

The Fat Olive

The Purple Onion

Kartel酒吧

yuccA
YUCCA餐厅

项目名称：*YUCCA*

坐落地点：上海思南公馆26F

设计公司：*Dariel & Arfeuillere*

——*A Lime 388 Company*

设计师：*Thomas DARIEL & Benoit ARFEUILLERE*

主要材料：喷绘地砖、壁纸、黑色油漆

摄影师：*Derryck Menere*

Yucca坐落在思南公馆26楼，在上海老法租界的一幢醒目的传统风格的建筑中。Yucca 是一家时髦的以现代摩登氛围和抓人眼球的设计装修为特色的墨西哥餐厅。

"墨西哥"这个词和餐厅、酒吧联系起来总会让人想到 一系列固定的套路：仙人掌，宽檐帽，插袋手枪，子弹带，暴露出人造砖块的裂开的灰泥墙，或者为了所谓的"上档次"而刻意放置的Frida Kahlo肖像画。本案设计风格区别于一般的套路， 设计出一种新的视觉享受。

Yucca is located in the Sinan mansion 26F, the old French Concession in Shanghai, an eye-catching building in the traditional style. Yucca is a chic modern atmosphere and modern design caught the eye decoration featuring Mexican restaurant.

"Mexico," the word and restaurants, bars, always brings to mind a series of linked fixed routine: cactus, wide-brimmed hat, pocket pistol, ammunition belt, exposed man-made bricks cracked plaster wall or to the so-called "upgrading" and can be placed portrait of Frida Kahlo. Case design is different from the general routine, designed a new visual experience.

时髦拉丁风，Yucca时髦华丽的室内设计以丰富的有冲击力的色彩搭配为特色。墙面用生动的蓝色和粉红色粉刷而成，地板使用了蓝白相间的几何图案的马赛克。一幅蓝色背景的女人照片从入口一直延伸到酒吧顶层，旋转楼梯引向顶楼私密空间。刻有警句"DESEO FUERZA AMOR LUJURIA"的大理石吧台体现了空间的精神。色彩的搭配，地板的图案，自然光的使用等各元素使来宾在酒吧这个空间里更加互动。

Fashionable Latin style, Yucca stylish interior design with gorgeous rich have more impact with color features. Walls with vivid blue and pink paint from the floor using a blue and white geometric pattern mosaic. A blue background woman photo has been extended from the entrance to the bar top, spiral staircase towards the top floor of private space. Engraved with the aphorism "DESEO FUERZA AMOR LUJURIA" marble bar embodies the spirit of space. Color matching, floor patterns, the use of natural light and other elements of the guests in the bar this space more interactive.

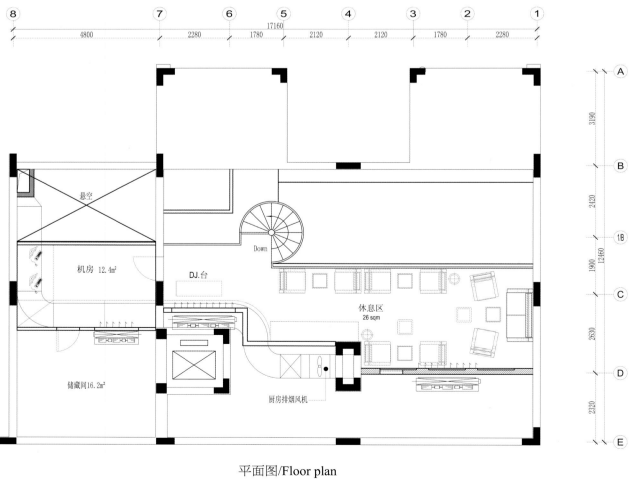

平面图/Floor plan

机房 12.4㎡

DJ.台

休息区
26 sqm

储藏间16.2㎡

厨房排烟风机

悬空

Down

"我们希望打造一个能激发灵感的场所，在这里，想像力可以自由驰骋，朋友们也可以互相寻找灵感"。Yucca的迷人，现代的墨西哥感觉来自于 Salvador Dali, Diego Rivera and Frida Kahlo的灵感激发。

"We want to create a place for inspiration, where imagination can run free, and friends can find each other inspiration." Yucca charming, modern Mexican feel from Salvador Dali, Diego Rivera and Frida Kahlo's inspired.

其宗教基调，西班牙样式的定制马赛克，鲜明不拘一格的蓝色，枝状大烛台，天主教十字架，精制地板，螺旋纹手扶椅、铁门，一系列元素打造了一个丰富热烈又温暖，舒适，和谐的Yucca。

Its religious tone, the Spanish style custom mosaic, distinctive eclectic blue, branched candelabra, Catholic crosses, refined floors, spiral armchair, iron doors, a series of elements to create a rich and intense, warm, comfortable harmonious Yucca.

山川设计

山川设计事务所1979年创立于香港，1997年成立上海公司，2002年成立北京公司。

山川设计自创建以来，一直将法、势、术三个要点作为好作品的工作引导。要成功发展一项事业，除了要掌握天时、地利、人和外，也要将法、势、术三方面不断互动发挥及影响。适当调整、平衡各因素，互助互推的发展。

Hotpot Shanghai

上海鲍鱼火锅餐厅

项目名称：鲍鱼火锅

坐落地点：上海市长宁区剑河路2378号（近虹桥路）

设计公司：山川设计事务所

主要材料：黑金沙、板岩石、马来漆、乳胶漆

ArtDeco最美的姿态就是承载着历史的璀璨瞬间，优雅华贵。鲍鱼火锅，美食中的贵族，选料考究，极致奢华。一如ArtDeco萦绕不散的高贵气质，艳翠紫椿，浑然天成。当ArtDeco端庄而曼妙的融入到空间的每一个角落时，鲍鱼火锅这传统的中式贵族也就充满了西式的浪漫优雅。

ArtDeco beautiful gesture that carries the history of the bright moments, elegant luxury. Abalone pot, food of the aristocracy, choice of materials known, the ultimate luxury. As ArtDeco lingering ghost of the noble qualities, Yan Cui Zi Chun, totally natural. When ArtDeco dignified and graceful into every corner of space, the abalone traditional Chinese hot pot which also is full of Western-style aristocratic romantic elegance.

空间设计意念：深邃优雅的时尚，含蓄中风华绝代。简约摩登的贵族时尚，复古与现代的融合，含蓄而优雅。光影与空间流动的对话，不张扬中蕴涵高贵品质，儒雅尊贵，雍容大度。藏在细节深处的美，是生活质量永远的追求。而疏离隔秀的空间所呈现的细节完美主义则是一种内心深处的情怀。

Space design concept: profound and elegant fashion, implicit in the absolute beauty. Simple modern aristocratic fashion, fusion of retro and modern, subtle and elegant. Lighting and space flow of dialogue, not play in the implication of noble quality, refined noble, generous. Hidden in the details of the depths of the United States, is the pursuit of quality of life forever. And retiring the space show presented every detail is a perfectionist feelings deep inside.

家具风格定位：似水年华，万千风情，典雅华贵。家具华贵的感觉都在细节上完美的体现，所有流畅的线条，优美的图案都浓缩着快乐，祥和的生命力。ArtDeco经过岁月的流逝，突现出特殊的色泽，令人惊艳，尽显精致、成熟的都会风华。

Furniture Style: various styles, elegant and luxurious. The feeling of luxury furniture are the perfect embodiment of the details, all smooth lines, beautiful patterns are concentrated with the happy and peaceful vitality. ArtDeco after the passage of time, the fore a special color, stunning, full delicate, mature and cosmopolitan elegance.

色彩和配饰设计意念：声色光影，曼妙如歌。温润的咖啡色系列，温和历练不失优雅，加上水晶灯璀璨的晶莹，简洁高雅的风情，让人怦然心动。每一个视觉的焦点，垂幔的天花，熠熠的火炉，精美的艺术品，总在不经意中泛出种种内敛的风情，莺飞草长，遍地花香。

Color and accessories design ideas: sensual lighting, as wonderful as the song. Warm brown series, moderate experience yet elegant, with bright crystal chandeliers, simple and elegant style, people can resist. Each visual focus, vertical mantle ceilings, gleaming stove, fine art, always casual Jingyi restrained style in all the glow, birds are flying long grass, flowers everywhere.

姜湘岳

中国十佳酒店设计师；国际商业美术设计师专业A
级资质、高级工程师资格、高级室内建筑师资格。

2001年至今，江苏省海岳酒店设计顾问有限公司
（原江苏省海岳建筑装饰工程设计有限公司）总
设计师

近期主要荣膺

"金羊奖—2008年度中国十大商业空间设计师"

获"第七届中国国际室内设计双年展"--银奖

"2008润澳星空间杯江苏省室内设计大奖赛"商
业空间类--大奖

"2008江南之韵室内设计大奖赛"--商业空间工
程实例类二等奖

"2009 INTERIOR DESIGN CHINA酒店设计
奖"--优秀奖

"2009年度中国饭店业设计装饰大赛-金堂奖"
酒店餐厅类--银奖

"2009年度中国饭店业设计装饰大赛-金堂奖"--
中国十大餐厅空间设计师

CIID "1989—2009中国室内设计二十年"--优秀
设计师

"2010 INTERIOR DESIGN CHINA酒店设计
奖"--优秀奖

"2010金外滩奖"—最佳酒店设计奖（优秀）

中国建筑装饰协会"全国有成就的资深室内建筑
师"称号

"2010中国国际空间环境艺术设计大赛"--酒店
空间工程类"筑巢奖"银奖

"2010亚太室内设计双年大奖赛"—酒店空间设
计优秀奖

Zi House Deluxe Restaurant
南京紫轩餐饮会所

坐落地点：南京市紫峰大厦

面积：*1550m²*

设计师：姜湘岳

主要材料：黑高光木饰面、丹麦灰镜、丝光布、意
大利黑金花石材

摄影：潘宇峰

本案位于紫峰大厦4层，定位为服务精英人士，是南京餐饮业奇葩、绿地广场的美食地标。

基于这一定位理念，所以从设计之初我们就抛开了传统思想，而是采取了一种融合现代主义和新古典主义风格的设计创想，力求让人们沉浸在优雅的文化氛围中亦能品味到现代时尚气息，即古典式现代美学之设计理念。

The case in the purple peaks Bldg, positioned as a service elite, wonderful restaurants in Nanjing, green square dining landmark.

 Based on this positioning concept, so we set aside from the initial design of the traditional thinking, but take a fusion of modern and neo-classical style to design creative, to let people immersed in the cultural atmosphere of elegance also taste the flavor of modern fashion, that is classic modern aesthetic of the design.

万宏伟

1991年毕业于四川美术学院
1992年开始从事室内设计
现任宁波市汉文设计设计总监
中国建筑学会室内设计分会会员

近期主要作品

宁波老外滩紫金会所
宁波醉美时尚中国菜馆
彩虹国际样板房系列
中建房产新纪元置业办公楼
聚卿坊私人会馆
博洋SI专卖系统
宁波国际交流服务中心
汉文设计办公室

WenDing No.1

文鼎一号

坐落地点：宁波市和义大道
面积：**2500m²**
设计公司：宁波汉文设计
设计师：万宏伟、胡达维
主要材料：金蜘蛛石材、蓝宝石石材、柚木饰面、
钢板刻画、手绘刺绣、手绘壁、香槟金镜子

餐厅位于宁波最为繁华的和义大道，地理位置十分特殊。建筑为橄榄形，因而重塑室内空间平面与功能成为重点，优质的主材确定及艺术风格马赛克拼花图形的多方位使用成为整个空间设计的主要元素，整体而优雅。手绘壁纸及订制灯具更为空间带来浓厚的艺术气息。

设计重点突出现代商务时尚餐饮空间与周边高端国际品牌的呼应与融合，更具国际的观感来表述设计主题。我们试着解放空间，体会宁静致远的意味。

Restaurant located in Ningbo, the most prosperous and Yi Road, location is very special. Building for the olive, which reshape the interior space and features a flat key, high-quality and artistic style of the main material to determine the mosaic using a multi-faceted graphic design of the main elements of the space, as a whole and elegant. Hand-painted wallpaper and custom lighting more space to bring strong art.

Design of modern business focused around high-end fashion and dining room echoed with harmonious international brands, to express a more international look and feel design theme. We try to liberate space, experience quiet attitude of the mean.

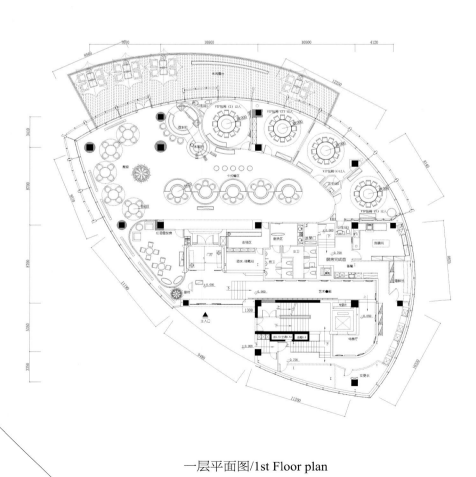

一层平面图/1st Floor plan

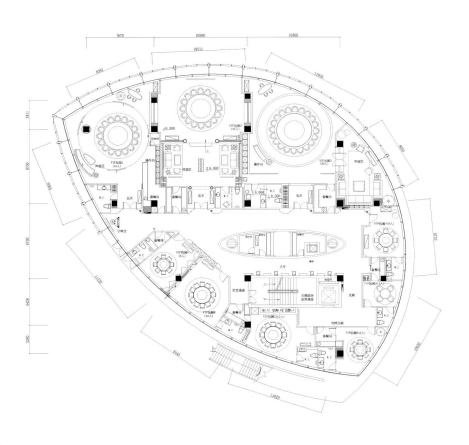

二层平面图/2nd Floor plan

利旭恒

英国伦敦艺术大学 BA(Hons) 荣誉学士

长年致力于酒店餐饮空间与商业地产的设计工作，10多年的酒店餐饮空间设计，累积了丰富的项目经验，在多个品牌设计风格与商业型态上都卓有成就。

近期主要作品

鼎鼎香

麻辣诱惑

港丽餐厅

烹大师烧肉达人

西堤牛排

王品牛排

豆捞坊

汉拿山

权金城

瑞贝卡

Matsumoto Japanese Restaurant
松本楼

坐落地点：北京国瑞城

面积：**650m²**

设计公司：古鲁奇公司

设计师：利旭恒、季雯

主要材料：灰砖、麻绳、木雕、铁艺

摄影：孙翔宇

松本楼日式烧肉隶属于国内大型餐饮集团旗下，是一家高端日式烧肉餐厅，以高级日本料理，高端日式烧肉为主。北京国瑞城的松本楼，整体装修风格以暗色为主，散客区中心是酒水吧和烧烤台。穿过散客区是隔出来的包房，温暖的黄色灯光让整个聚会变得私密又舒服。

Matsumoto Japanese Restaurant Group under the domestic large-scale catering, pork is a high-end Japanese restaurant, with high Japanese cuisine, high-end Japanese-style pork-based. Beijing Glory City of Matsumoto Japanese Restaurant, the whole decorated with dark-based, individual district is wine bar and barbecue sets. Across the hall is separated out of the rooms, the warm yellow light for the whole party to become intimate and comfortable.

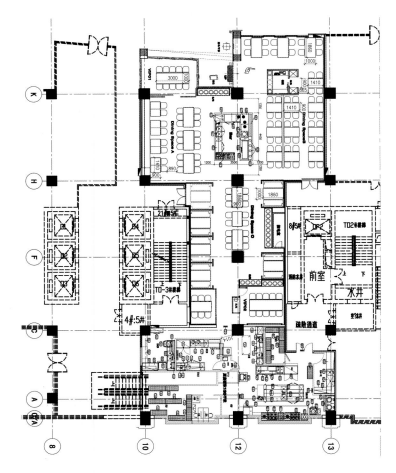

平面图/Floor plan

利旭恒

英国伦敦艺术大学 **BA (Hons)** 荣誉学士

长年致力于酒店餐饮空间与商业地产的设计工作，10多年的酒店餐饮空间设计，累积了丰富的项目经验，在多个品牌设计风格与商业型态上都卓有成就。

近期主要作品

鼎鼎香

麻辣诱惑

港丽餐厅

烹大师烧肉达人

西堤牛排

王品牛排

豆捞坊

汉拿山

权金城

瑞贝卡

Hannashan Korean BBQ

汉拿山北京新源里店

坐落地点：北京朝阳区新源里

面积：**1500 m²**

设计师：利旭恒

主要材料：花岗石、济州岛火山洞石、铁锈板、锡箔纸、橡木、**LED**光源

摄影：孙翔宇

本案设计师在空间操作手法上着重于不等比重的空间分割概念，一座看似复杂且层层交错的空间主体结构，透过前后三个独立用餐区，挑高的空间玻璃墙面与巨大的造型灯笼导入中式韩风的设计概念主题，清晰地引导出这座城堡的空间动线。

用餐区整体空间是由一个个大小相同但是样式不统一的座椅构成，前区用餐区的卡座椅与上方的白色大圆灯笼交相呼应，同时白色大圆灯笼中和了浓郁的烧肉口味，在给人们带来味蕾的满足之时，用视觉的方法来舒缓桌上炭火的炙热触感。

Case operating practices in space on the designer to focus on the concept of varying the proportion of the space division, a seemingly complex and the main structure of the space layers staggered through the front and rear three separate dining area, high-ceilinged space with a huge glass wall Famous Chinese lanterns into the shape of the design concept of theme, a clear space to guide the moving castle of the line.

Dining area as a whole space is the same size, but by a pattern of non-uniform composition seat, front seat area and dining area of the card above the large white cross echoes garden lanterns, garden lanterns in both large white and the rich pork taste, bring to people the satisfaction of the taste buds, the use of visual methods to relieve the hot charcoal touch the table.

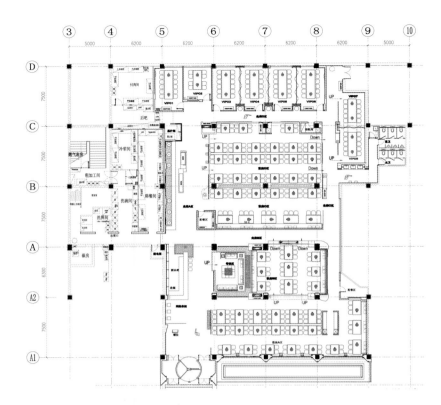

平面图/Floor plan

陈旭东

吉林艺术学院设计学院 教授

北京浩昌工程有限公司 设计总监

美国**RB**建筑设计事务所 建筑师

中国建筑学会室内设计分会 会员

近期主要作品

2005年，第二届海峡两岸室内设计大赛《凝固的意象》获办公类一等奖；《儒雅与时尚的私语》获酒店类二等奖；《湖边的阳光屋》获居住类三等奖

2005年，第一届国际室内设计大赛及"华耐杯"中国室内设计大奖赛《凝固的意象》获商业工程类三等奖；《东方饺子王》获酒店工程类佳作奖

2005年，吉林省第五届建筑画展，建筑画作品《别墅设计》获一等奖；《艺术中心》获二等奖

2005年，省级优秀勘察设计，建筑设计作品《吉林艺术学院宽平校区教学楼》获一等奖

2006年，第二届国际室内设计大赛及"华耐杯"中国室内设计大奖赛《晨光文苑售楼处》获工程类三等奖；《咖啡小镇售楼吧》获方案类二等奖

周玉玮

吉林艺术学院传媒学院

2008年7月至今，陈旭东设计事务所工作

Oriental Hotpot
东方火锅（中东大市场店）

项目名称：东方火锅（中东大市场店）

面积：*1000 m²*

设计师：陈旭东、周玉玮、刘洋

主要材料：白色直纹大理石、木纹防火板、黑金沙大理石、灰镜片、钢化玻璃

摄影：陈旭东

东方火锅中东大市场分店坐落于长春市自由大路与东环城路交汇处。一般而言，开火锅都要求体现"火"的热情与"火"的内涵相关联，而本案的设计却恰恰相反的采用了水的属性特征，水与火在此相融相生。室内墙面与地面大面积采用了蓝灰条相间的白色直纹大理石，其肌理特征很像水或冰层的效果。

Middle East hot pot large stores located in Changchun City, the market freedom of the road with the East Ring Road Interchange. In general, open pot are required to reflect the "fire" of passion and "fire" associated with the content, while the opposite case, the adoption of the design of the properties of water features, water and fire in this integration. Interior walls and floor of a large area with a blue-gray and white marble, white ruled, and its texture characteristics like water or ice effects.

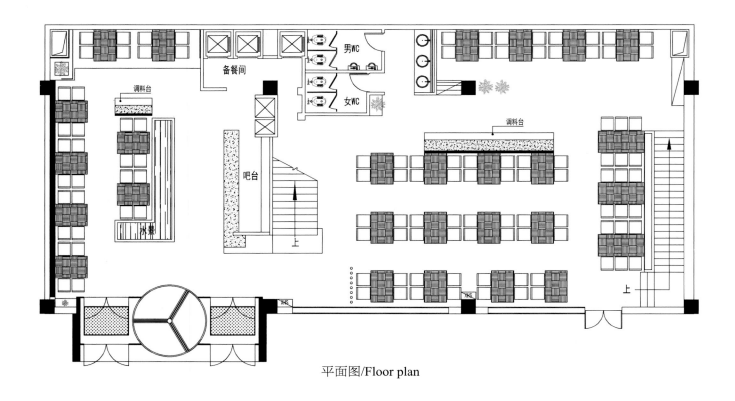

平面图/Floor plan

"火"的意喻是通过红色的木板及红色座椅以及红色窗帘的运用来加以强调。同时也采用了不锈钢片组合构成火焰的形态装饰室内。使得"火"在材料的特性上自然的向"水"的特性上加以转换，从而使"火"与"水"意喻相融。

"Fire" analogy is the red wood and red seats and red curtains to be emphasized that the use of the expression of the fire. Also uses a combination of stainless steel sheet to constitute a form of decorative indoor flame. Making the "fire" in the properties of the materials on the nature of the "water" of the characteristics to be converted, so that the "fire" and "water" meaning integration.

王仲文

中国建筑学会室内设计分会会员 室内建筑师

1997年，毕业于哈尔滨轻工业学院室内设计系

1998年，任哈尔滨市装饰有限公司设计部主任

2002年至今，任哈尔滨唯美源装饰设计有限公司
创意室主任兼设计师

近期主要作品

哈尔滨光谱美容SPA馆

哈尔滨惠鸿美容SPA馆

北京御香苑肥牛府

哈尔滨国际饭店

哈尔滨梦天湖韩式松骨会馆

哈尔滨采珍集火锅

哈尔滨SOKO健身会所

哈尔滨信恒会所

伊春小兴安岭恐龙博物馆

北京名华四季洗浴会馆

HuaShanHui Restaurant

华膳汇

项目面积：**3550m²**

项目地点：哈尔滨市南岗区西大直街

设计公司：哈尔滨唯美源装饰设计有限公司

主要材料：橡木板、理石、皮革、不锈钢

摄影：张奇永

砌一壶清茶，漫步在这幽静的环境中，犹如太湖一般的幽静，置身世外桃源的感觉，屏住呼吸享受着大自然的芬芳……

曾几何时，人文竟成为人们苦苦追寻的一种时尚，历史已成为过去，怀旧不是时空逆转，而是打破传统文化在人们心中的固有形象，用全新的理念从新诠释古典文化成为必然。

Assemble a pot of tea, walk in this quiet environment, like a quiet lake in general, exposure to the feeling of paradise, hold your breath to enjoy the fragrance of nature ... ...

Once upon a time, the humanities have become a people struggling to pursue fashion, history of the past, nostalgia is not time reversal, to break the traditional culture inherent in people's minds the image, with new ideas from the new interpretation of classical culture became a necessity.

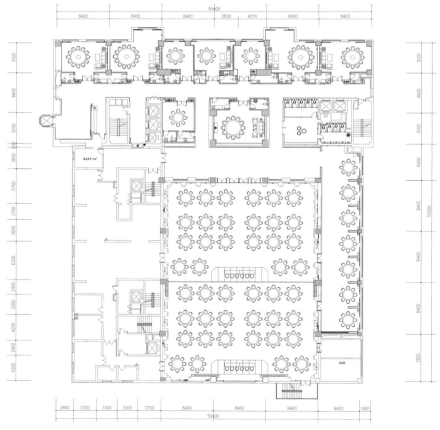

平面图/Floor plan

本工程的造型独特别出心裁，各种奇异的镂空造型在灯光的映衬下格外耀眼，复古的木质窗户更加衬托出这里的幽静。石材墙面上刻有竹子的纹样，让人们不由地联想到竹林的幽静。

Project unique shape fantastic variety of exotic hollow shape especially in the light against the background of bright, retro wooden windows bring out here is quiet. Stone wall engraved with bamboo patterns, so that people could not help think of bamboo quiet.

恰当运用石头、木制品、竹子等传统元素，使本案设计充满了文化气息，太湖石的抽象形态结合中国烟云的气态，彰显出东方的神韵。充分的利用大堂在整个建筑的位置，让这里得到最大的视线和最好的风景，同时利用具有浓重的文化特色的主背景，身处在其中，有一种历史与自然融为一体的感觉。

Appropriate use of stone, wood, bamboo and other traditional elements of the case design is full of cultural atmosphere, combined with China's Taihu abstract form gaseous clouds, sheets showing oriental charm. Full use of the location in the building lobby, and make it get the maximum attention and the best scenery, while taking advantage of a strong cultural characteristics as the main background, living in which there is a sense of history with nature.

王砚晨

毕业于西安美术学院
意大利米兰理工大学 国际室内设计硕士

经典国际设计机构(亚洲)有限公司 首席设计总监
北京至尚经典装饰设计有限公司 首席设计总监
中国建筑学会室内设计分会 会员

李向宁

意大利米兰理工大学 国际室内设计硕士

经典国际设计机构(亚洲)有限公司 艺术指导
北京至尚经典装饰设计有限公司 艺术指导
中国建筑学会室内设计分会 会员

Meizhou Dongpo Restaurant

眉州东坡酒楼国奥村店

面积：5000m²
坐落地点：北京市朝阳区林萃东路2号院甲3号楼国奥中心A座1层
设计：王砚晨、李向宁
主要材料：回收材料（可乐瓶、易拉罐、红酒塞、铁钉）、牙签、跆拳道腰带、毛笔、蜀绣、竹编

2008北京奥运已成为历史，但它带给国人的激情与梦想，自豪和自尊却依然在延续。当眉州东坡酒楼奥运主题餐厅在国奥村落成之际，我们心中仿佛又重新点燃那已熄灭的奥运圣火。

整个餐厅以2008北京奥运为主题，试图用艺术化的手段重新演绎奥运的主题与精神。"同一个世界，同一个梦想""北京欢迎你"这些2008年耳熟能详的口号，今天成为设计中的主线，让我们在后奥运时代依然感受到奥运精神传承的脉搏。

2008 Beijing Olympic Games has become history, but it brings people of passion and dreams, pride and self-esteem still continues. When the occasion of the completion of the Olympic theme restaurants, our hearts seem to have re-ignite the Olympic flame that has been extinguished.

 The Olympic-themed restaurant in Beijing, trying to use the means of artistic re-interpretation of the theme and spirit of the Olympics. "One World, One Dream," "Beijing welcomes you" these slogans, the design of today become the main line, so we still feel in the post-Olympic era, the pulse of Olympic legacy.

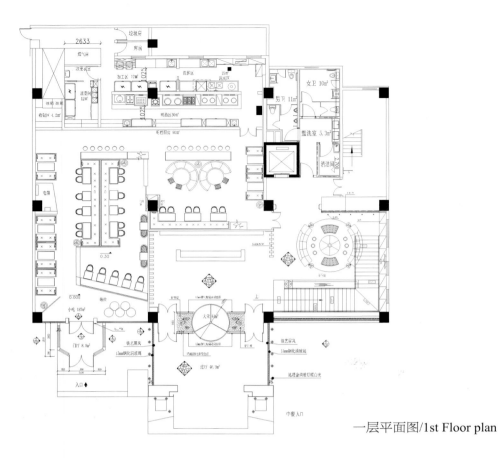

一层平面图/1st Floor plan

"有朋自远方来，不亦乐乎"，2008奥运会开幕式上的一幕幕精彩画面，仿佛瞬间凝固定格在空间中。奥运火炬传递的城市线路在地面上延伸，光电感应的互动让人们体验到了火炬传递的激情与艰辛，而大门上近乎逼真的火炬拉手，也让更多的人圆了火炬手之梦。

"We are very pleased to have friends from afar", the 2008 Olympic Games opening ceremony of the wonderful scenes on the screen, like an instant freezing freeze in space. Olympic torch relay city line extension on the ground, the interaction optical sensors to allow people to experience the passion and hard to pass the torch, and the door handle on the torch close to realistic, but also allow more people to round up the torch of the dream.

极具中国神韵的二十八个运动项目的小篆图形，分布在包间中，每一个包间是一个运动主题，传统的小篆被赋予了新的视觉语言，艺术装置的表现手法用来重新演绎不同运动项目的特点。秉承绿色奥运的理念，大量的使用餐饮经营中的回收材料，将绿色奥运的精神理念用艺术化的手法传递给大众。

Great charm of the graphics China, distributed in rooms, each private room is a campaign theme, traditional Xiaozhuan been given a new visual language, art device performance techniques used to re-interpretation of the characteristics of different sports. Adhering to the concept of Green Olympics, a lot of use of recycled materials in the food business, the spirit of the Green Olympics concept of using artistic means to pass to the public.

二层的奥运记忆长廊集中呈现了眉州东坡在2008奥运会供餐期间留存的珍贵历史资料——胸牌，车证、工作证、员工日记、照片、开闭幕式菜单等，这些生动形象的物品以博物馆的陈列方式展示出来，见证了眉州东坡在奥运供餐工作中的点点滴滴。

Concentration of the second floor gallery presents the Olympic memories restaurant meals in the Olympic history during the retention of valuable information "badge, car cards, work permits, staff diaries, photographs, opening and closing ceremonies menu", etc. These items display in the museum's way of showing out to witness the restaurant in the Olympic feeding work everything.

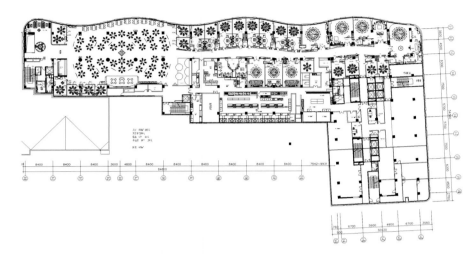

二层平面图/2nd Floor plan　　　　　　　　　三层平面图/3rd Floor plan

三层的整体风格呈现出传统的北京风貌，古朴的青砖，精美的汉白玉门墩，细腻的古铜门楣在空间中交相辉映，同那些老北京胡同中随风摇曳的树枝，飞翔的小鸟，组成一幅和谐的画面，体现出奥运带给古老北京的新面貌。

The overall style of showing a three-tier traditional Beijing style, rustic brick, beautiful white marble door stones, delicate bronze lintels each other in space, with those old Beijing Hutong in swaying branches, flying birds, form a harmonious picture, reflecting the ancient Beijing Olympics to bring a new look.

甘泰来

纽约哥伦比亚大学 建筑硕士（荣誉获奖毕业）

纽约康乃尔大学建筑与都市设计研究所

东海大学 建筑学士

齐物设计事业有限公司 总监

交通大学建筑研究所 讲师

实践大学建筑研究所 讲师

Brown Sugar

Brown Sugar 台北店

设计单位：齐物设计

设 计：甘泰来

建筑面积：室内198 m²、室外549 m²

主要材料：和平白石材、橡木染深色、镜
面不锈钢、灰色洗石子

摄 影：卢震宇

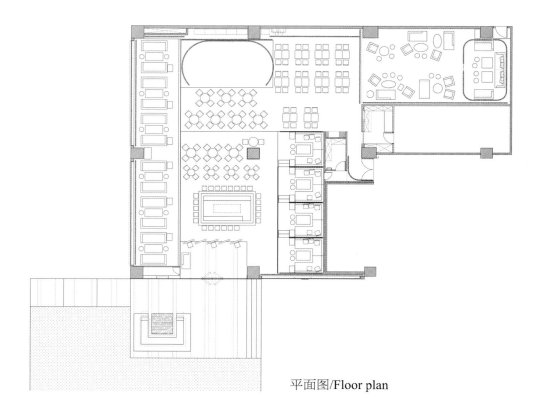

平面图/Floor plan

本案在设计上承袭了上海店"阶梯剧场"的概念，在开放空间内规划多种平台高度，共划分为吧台区、沙发区、阶梯包厢区、散桌区、舞台区、VIP区，基地前端为架高吧台区与尽头舞台区中包夹着散桌区，左右两侧为架高的沙发区与阶梯式的座位。

The case in the design of a "ladder theater" concept, open space planning in a variety of platform height, is divided into the bar area, sofa area, step box area, stage area, VIP area, the base front-end for the elevated bar area and at the end of package clamped in the stage area tables scattered areas, right and left sides of the sofa for the elevated seating area and ladder.

在刻意放宽的沙发区走道与家具不贴墙的摆设方式下，使得动在线更为自由。以不锈钢材质作为人字纹屏风以及立面、地台的框饰，并大量使用深色镜面及茶色玻璃，使空间中镜射出霓虹光彩的华丽幻影。

Designer sofa area in the aisle of furnishings and furniture are not wall-mode, making moving online more freely. Separated by stainless steel as well as across the facade, decorated floor of the box, and extensive use of mirror and dark tinted glass, so space mirrors the gorgeous neon injection.

il buco delicatessen+restaurant

il buco 熟食店

坐落地点：18, Sarri Str and 2, Sachtouri Str, Psirri,
GR 105 53, Athens Greece
设计单位：Dionisis Sotovikis and Kirki Mariolopoulou
设计：workshop dionisis+kirki and Sotovikis+B
建筑面积：123 m²
摄影：Vassilis Makris

IL buco 熟食店位于15sq.m的街角商店，是该地区的主要街道营业场所之一，其设计概念是把它作为一个小盒子。橱窗向外开启，你能看到可以品尝美食。熟食店是如此之小，其中的设计体现了建筑师的"洞"的概念，这是名字来于"il buco"是意大利语中的"洞"。熟食店周围发生的一切事情都能记录在外墙上，这就是为什么建筑师使用胶合板，这种材料无论是用于涂鸦还是用于张贴画，它都将保留其本身的特性。

Il buco_deli is just a 15sq.m corner shop and the concept was to treat it as a small box that opens up and closes on one of the main streets of the area. The benches open outwards inviting you to see and taste what's inside. The deli was painted black so when closed it vanishes and becomes one with the rest of the street's buildings. The deli was so small, that instead of a shop window or show case, the architects enhanced the notion of the "hole", and that is where the name came from: "il buco" means "hole" in Italian. Everything that happens around the deli is to be recorded on its facades and that is why the architects used plywood, a material that even if it is used for graffiti or billposting it will retain part of its character.

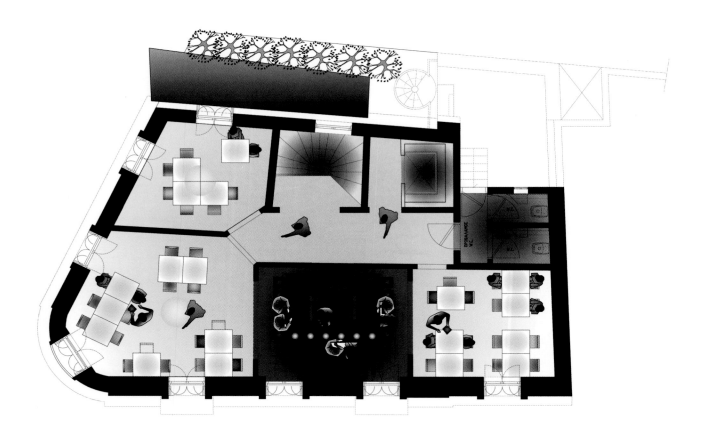

平面图/Floor plan

IL buco熟食店为餐厅的设计提供了设计的灵感，餐厅位于马路对面的有着新古典主义风格建筑的三楼。餐厅有三个房间，在餐厅的中央，是一个鸡尾酒酒吧。这个有着黑色胶合板货架的鸡尾酒酒吧成为实现区分不同的空间，连接熟食店和餐馆之间纽带。

Il buco deli was the inspiration for Il buco_restaurant that was designed at a later stage, on the third floor of the neoclassical building across the street. The restaurant has three rooms around the central space that is the cocktail bar which also operates as a mini-deli. In order to create a connection between the deli on the street and the restaurant, but also to emphasize the different mood of this space to the rest of the restaurant rooms, the cocktail bar was also painted black with plywood shelves.

其余房间所有的墙壁和地板被漆成白色显得更明亮。黑色铁腿的纯白色桌子和旧木椅子多来自于古董店。其中，还有些比较新的带有藤木编织的座椅。餐厅中间85盏吊灯，营造了华丽的空间氛围，在设计中采用"少即是多"的设计原则。此外，古董橱柜和大理石顶部的组合搭配有力的补充空间。墙壁上悬挂的艺术作品为描图纸制作的。这间餐厅给每位前来用餐的客人，营造了一种温暖而怀旧的空间。

The rest of the rooms are much brighter with all the walls and floor painted white. The plain white tables with black iron centered legs are complimented with old wooden chairs that were bought from one of the many antiques' shops of the area of Psirri. Some of them were given a fresher look with new seats made from woven leather strips. The low ambient light of Droog's 85 lamps by Rody Graumans has been used to light up the space and create an opulent chandelier, where 'less and more' are united in a single product. Furthermore, an antique sideboard with a marble top 'flirts' with the mix and match antique chairs and is complemented by Red Revisited plate of Droog. On the walls hang selected art works of Dionisis Sotovikis made from tracing paper, water and fire. This play between the old and the new leaves visitors with a warm nostalgic feel, which is very characteristic of the local atmosphere.

周易

周易室内设计工作室 主创

周易概念建筑工作室 主创

1979年，以自学方式学习建筑与室内设计

1989年，创设JOY室内设计工作室

1995年，创设JOY概念建筑工作室

近期主要荣膺

2009年，**TID AWARD** 商业空间设计大奖——庵锅

2009年，商业空间设计入围——轻井泽员林店、客家本色餐厅公益店

2010年，**TID AWARD** 商业空间设计入围——轻井泽文心店、客家本色餐厅朝富店

Sanyi Mountain TaohuaYuan

三义山中桃花源

建筑面积：*2280.1m²*

坐落地点：台湾三义

设计公司：周易设计工作室

设计师：周易

主要材料：清水模、铁件、玻璃、铁刀木

摄影：和风摄影

山中桃花源以简约、素朴的姿态静静地倚偎在山丘上，不论是从主入口或停车场入口进入，迂回绕进建筑物内部空间时，都不禁会有豁然开朗的感受。设计师意图为现代人展开一页新桃花源记，采用混凝土墙面、水波、光影、布幔以呼应区域环境特色的客家风景。进入内部空间后，首先见到大面水池，水池外围垂吊长形红色布幔，摇曳着一股闲适自得的气氛；远处的混凝土墙，则因着光影点缀，丰富了空间表情。

Restaurant with simple, naive attitude Yiwei quietly on a hill, whether from the main entrance or parking lot entrance, into the interior space, has suddenly see the light of experience. Designer intended to launch a new page for the modern Land of Peach Blossoms, with concrete walls, water, lighting, curtain and echo characteristics of Hakka regional environmental landscape. Into the internal space, first to see the Omo basin, basin-shaped peripheral long hanging red curtain, swaying with an atmosphere of quiet contentment; far from the concrete wall, due to the lighting decoration, rich spatial expression.

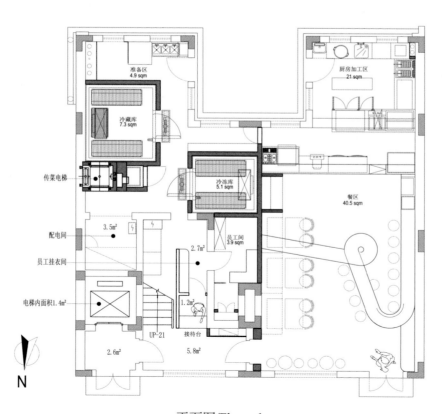

平面图/Floor plan

踏上水池中央的走道，走道上点点光亮，引着来客进入室内空间；设计师在一进门的地方放置了成对的红烛台，搭配壁面大幅传统牡丹花巾，成为室内空间的焦点。沿着阶梯往上来到二楼的餐饮区，设计师利用玻璃搭接围墙与主体建筑物，玻璃上的卵石则强化了光影表现；不规则的围墙作为餐饮空间背景，都掩映出丰富的光影表情。

Foot pool central aisle, the aisle a little bit on the light, cited the visitors into the interior space; designers put in a door where a pair of Candle sets with large traditional Peony towel wall, become the focus of interior space. Along the ladder up to the second floor dining area, designers use glass bonding with the main building walls, glass pebbles are enhanced lighting performance; irregular wall as a dining space background, have set off a rich expression of light and shadow .

混凝土的施工过程中，设计师使用木造模板，以达到混凝土壁面光滑平整的效果。建筑物整体透过混凝土的呈现与山间景致产生的和谐关系，展现出内部空间的结构，从而营造舒适的用餐空间。

Concrete construction process, designers use wooden template, in order to achieve the effect of smooth concrete wall. The overall presentation of the building through the concrete produced with mountain views and harmonious relationship, to show the internal space of embellishment emphasizes the Hakka style in the ground, hoping visitors will have a comfortable dining room, meals and local culture while also produced the dialogue.

Dariel & Arfeuillere

Thomas Dariel 和 Benoit Arfeuillere是总部位于上海的创意机构 Lime 388 的创立者

2010年，获得 "年度最具潜力设计师奖"

近期主要作品

The Funky Chicken

Yucca（金堂奖 2011 China-Designer 中国室内设计年度评选 年度十佳优秀餐饮空间设计）

12 Chairs

The Fat Olive

The Purple Onion

Kartel酒吧

The Purple Onion Restaurant
The PURPLE ONION餐厅

项目名称：*The PURPLE ONION RESTAURANT*

坐落地点：上海思南公馆26F

设计公司：*Dariel & Arfeuillere*

——*A Lime 388 Company*

设计师：*Thomas DARIEL & Benoit ARFEUILLERE*

主要材料：喷绘地砖、壁纸、黑色油漆

摄影师：*Derryck Menere*

The Purple Onion 坐落在老法租界。周边是令人愉悦的住宅及零售区域、中式和法式建筑、国际餐厅及小店、传统的弄堂，沿街绿树成荫……今日城市构成元素的完美结合。两位设计师Dariel & Arfeuillere 希望Purple Onion与这些元素互动，在这片颇具历史感的区域显得既传统亲切又卓尔不群。由世界知名大厨David Laris提出概念的The Purple Onion餐厅的设计灵感，由在南欧流行的颇具家庭温暖氛围的法式Bistrot及意大利式Trattoria激发。

Purple Onion is located in the former Shanghai French Concession. The neighborhood is a pleasant mix of residential and retail areas, Chinese and European style architecture, international restaurants and tiny stalls, wide tree lined streets and traditional lanes··· a perfect combination of what constitutes the city today. The two designers Thomas Dariel & Benoit Arfeuillere wanted with Purple Onion to play with these contrasts, to be traditional and familiar yet provoking and offbeat in this very heritage district.

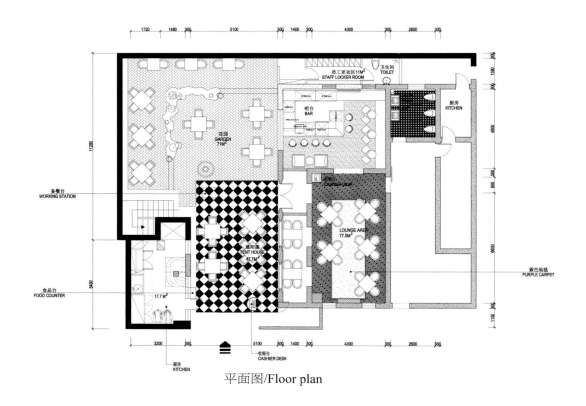

平面图/Floor plan

空间用三种颜色处理，黑色、紫色和白色。黑色，引领了餐厅的主要部分，结合家庭氛围的壁炉传达了私密、舒适的家庭感觉。黑色表达了潮流及优雅。在酒吧区，紫色为酒吧区特别打造的慷慨精致的色彩带来温暖、好客及迷人的氛围。白色中和并凸显了黑色紫色这两种主导色彩，垫子和光泽材料的使用也是这样。

The space has been treated around 3 colors, black, purple and white. The Black, prevailing the restaurant part, reveals the emotions and conveys a feeling of intimacy and comfort strengthened by the home-style fireplace. Foremost, the black color states for style and elegance. In the Lounge area, the Purple - a generous and sophisticated color specially created for the space - brings a warm, welcoming and glamorous atmosphere. Finally, the white gives a pretext to contrast and highlight the two predominant colors, as does the use of mat and glossy materials.

除了酒吧和餐区，The Purple Onion还有令人惊艳的带有林荫的庭院，隔开街道，让客人尽情享受老式建筑的外部风情。镶嵌在弄堂的中心，Purple Onion隔离了这喧嚣城市中的疯狂，像一座宁静的岛屿，给人以家的温暖，同时又非常时髦迷人。整体设计通过体现其文化、符号和性格而赋予了The Purple Onion强烈的可识别性，从而将其打造成极具个性的独特空间。极具个性的美味餐点由天才厨师Lex Hauser主理。

Beside the bar and fine dining areas, Purple Onion also benefits from a stunning tree-shaded courtyard protected from the street that gives ones the opportunity to enjoy the outside part of the old building. Nested in the heart of a small lane, Purple Onion is an intimate cocoon free from the city 's craziness, an island of peace where it feels like home, yet a trendy and glamorous one. Beyond the restaurant, the Design allows Purple Onion to become a whole small universe doted of a strong identity with its culture, symbols and characters.

高雄

资深室内设计师

中国建筑室内装饰协会 建筑室内设计师

中国建筑学会室内设计分会 会员

建筑装饰装修工程师

2002年，上海同济经典 主任设计师

2003年，福州闽辉设计工程有限公司 主任设计师

2004年，福州缔创设计顾问有限公司 设计总监

现任国广一叶设计工程有限公司 副所长

Impression Wang Jiangnan Restaurant

印象望江南餐厅万达店

坐落地点：福州市仓山万达

面积：**550m²**

设计单位：道和设计机构

设计师：高雄

参与设计：吴运棕、高宪铭

主要材料：生态木、楼兰仿古砖、超白玻璃烤漆、

五厘银镜、乳化玻璃、黑白根、黑色不锈钢

摄影：周跃东

本案是道和设计在"望江南"系列餐厅设计中进行的一次不同尝试——带有新东方情节、江南韵味与现代抽象视觉感受的结合，整体设计呈现时尚、简洁的气质，黑白色相间、简约，线条具有一种直抵人心的力量。

黑白色反差、镜面反射的场景，具有强烈的舞台感。店内家具均以咖啡色绒布和大理石台面为主，墙面造型均以凹凸的绿可板及镜面为主，强烈的厚重感与跳跃简洁的线条给客人带来感观的冲击，也体现了餐厅的品牌定位。整体色彩对比强烈，但空间更富于流动和变化。

This case was, "Wang Jiangnan" series to a different restaurant designs in an attempt - with the new East plot, southern charm and modern abstract combination of visual experience, the overall design fashion show, simple temperament, black white and white, simple lines power.

Black and white contrast with a strong sense of the stage. Furniture store are mainly brown velvet and marble countertops, wall mirror shape are mainly laden with strong clean lines and jump to the guests to bring a sense of the concept of impact. Overall color contrast, but the space is more full of movement and change.

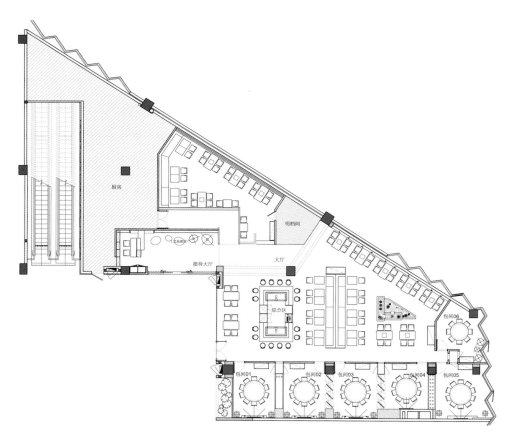

平面图/Floor plan

Shanghai-Style Restaurant
【海派】餐厅

项目名称："海派"餐厅
坐落地点：上海嵩山路
设计公司：SUPERPHTATO
摄影：申强

本案的餐厅融合了上海风情的精致和法式小馆的随性浪漫，充分体现了酒店所在的地域特色。餐厅露台可以尽览上海新天地的繁华景象，为食客提供亲切舒适的用餐环境。

The case of Shanghai-style restaurant combines fine and with the nature of romantic French Bistro, where the hotel is fully reflects the geographical characteristics. Shanghai Xintiandi restaurant terrace overlooking the downtown scene, to provide for the patrons warm and comfortable dining environment.

餐厅设有开放式厨房，大面积木块交错叠加的操作台如剧场般向食客展示制作过程，在你喝茶、看报的同时，也可以回到上海胡同的感觉。

Restaurant with an open kitchen, large pieces of wood spliced the console, such as theater-like display of the production process to diners in your tea, read newspapers, but also can go back to Shanghai alley feel.

餐厅在夹层设有四间私人包厢，设计师在这里阐述了历史与文化，风格与风情的微妙。镂空铸铁隔断，泛着皎洁灯光的黄锈石墙面，几株纤细的枫树，走道上几个参差摆放着的陶罐，如行云流水般的脱俗和清新。

In the mezzanine restaurant has four private dining rooms, where the designer describes the history and culture, style and subtle style. Hollow cast off, the lights glowing bright yellow rust stone walls, several strains of delicate maple, aisles stocked with a few mixed pottery, such as to fill the gap of the refined and clean.

田军

大连轻工学院家具设计专业

1997年成立田军设计工作室

2006年成立北京瑞普设计有限公司，2006年开始和俏江南餐饮有限公司合作至今，完成俏江南全国近30家的餐厅环境设计。

South Beauty—Suzhou Ligongdi

俏江南—苏州李公堤店

项目名称：俏江南-苏州李公堤店

项目地点：苏州

面积：*6000 m²*

设计单位：北京瑞普

设计师：田军

主要材料：窗格、皮革刺绣、理石、不锈钢荷花、喷绘、木制作

摄影：李为

江南名城苏州，有号称中国最大的内城湖泊——金鸡湖。俏江南苏州店就坐落于其中唯一的湖中长堤"李公堤"的岸边。此项目是一座建筑面积约6000平方米的独栋别墅。在完美的李公堤园区规划里，俏江南的位置极其优越，整个建筑依水而建，室内光线通透，以至于每个角落都能欣赏到旖旎而又壮丽的湖面风光。苏州城固有的文人气质，以及项目内部世外桃源一样的环境，都决定了室内装饰风格要将江南的诗情画意延续到底。

South Beauty Suzhou is located in one of the only shop on the lake in Long Beach "Ligongdi" the shore. Project is a construction area of 6000 sqm of single-family villas. In a perfect plan in Ligongdi Park, South Beauty's location is extremely favorable, the whole building was built by the water, indoor light, transparent, every corner of the lake can enjoy the magnificent scenery. Suzhou literati temperament, and beautiful environment within the project are determined southern interior poetic style.

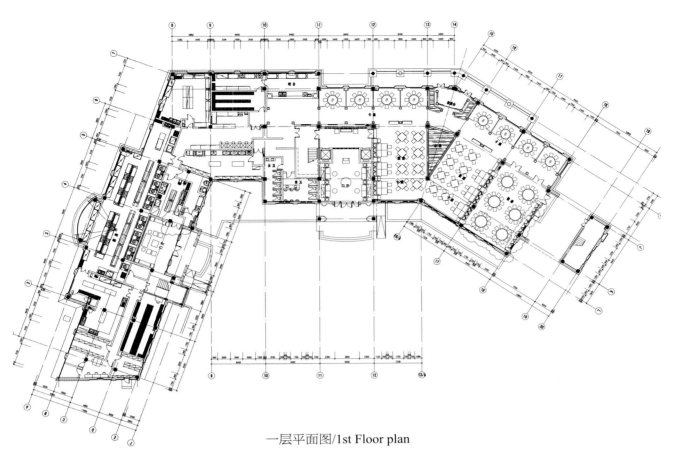

一层平面图/1st Floor plan

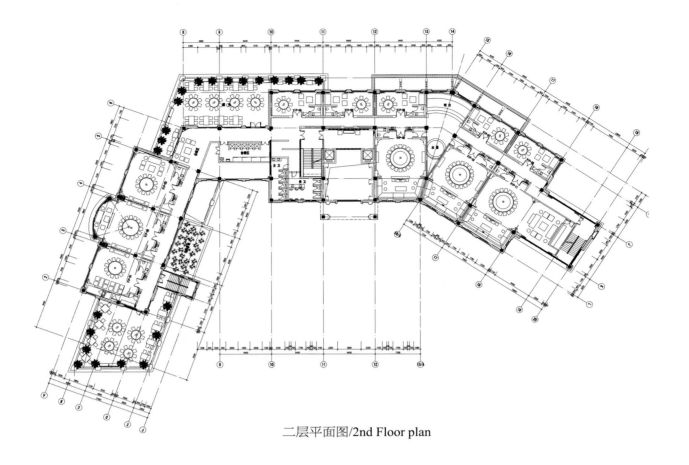

二层平面图/2nd Floor plan

虽然风格上我们依旧尊崇传统意象，但突破中国符号，挖掘延续古典之美，才是我们一贯秉承的理念。所以，在门厅入口，我们利用空间挑高的优势，做出了超高尺度的家具，用富有张力的手法，表现出迎接宾客的隆重。在过廊两侧及包间入口，我们加入了廊柱，红色丝绒的帘幕，葫芦装饰挂件等用来烘托气氛。散座及沙发卡座区，突破以往扇面为一个平面的表现手法，我们将它在立面与顶面上翻折，这样一来，坐落于其中的宾客，都会有温馨而有趣的奇妙体验。在空间对称布局的中央区域，设置了小桥流水，庭院表演，并特请艺术家在水池中央用金属质感的材料，量身定做了大片的荷叶、莲藕等雕塑，为周围的人们提供了一种传统与现代的穿越感。

We are still respected in the style of traditional imagery, but the breakthrough Chinese symbols, tap the classical beauty. In the entrance, the advantages of using high-ceilinged space, made a high-scale furniture, full of tension with the approach, showing the grand welcome guests. In the corridor, we joined the pillars; red curtain, pendants and other decorative gourd heighten the atmosphere. Loose seat and sofa area, breaking the previous expression, we will face it with the top surface of the turn-up, so is located in one of the guests, there will be a warm and interesting experience. Symmetrical layout in the central region of space, set up small bridges, garden show, and asked artists in the central pool of metal material and texture, tailored to the large lotus leaf, lotus and other sculptures, to provide people around the one, through a sense of tradition and modernity.

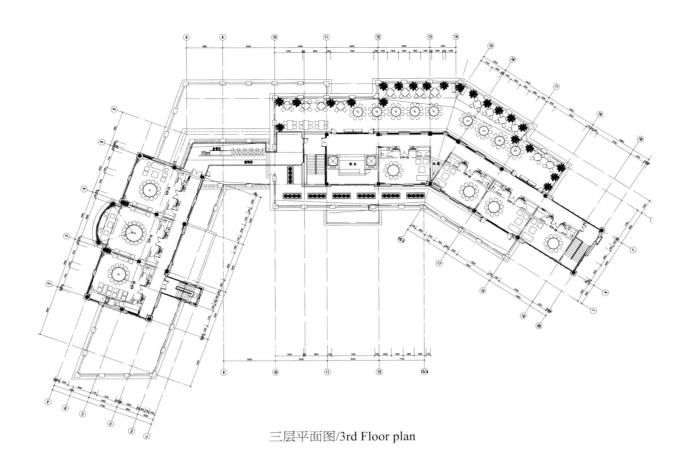

三层平面图/3rd Floor plan

徐岭啸

美国剑桥麻省理工大学 建筑学硕士

美国纽约哥伦比亚大学 土木工程学学士

ANS国际建筑设计与顾问有限公司 设计董事

美国LEED认证专家

近期主要作品

武汉东湖行政中心设计

上海庆安浦三路1229号综合体

天津住宅集团办公及检测中心

广州国会会所

上海思南公馆橘色火锅餐厅

上海思南公馆橘日本料理餐厅

上海Giuliano Fujiwara 旗舰店

汤昊隽

毕业于上海东华大学拉萨尔国际设计学院室内设计专业

近期主要作品

上海DNA办事处

上海米加餐厅

上海Fujiwara服装店

上海蓝蛙酒吧

上海兰博基尼展厅

上海Bekaert办公室

上海EMC办公室

上海外滩俱乐部

北京Suzie wong俱乐部

Japanese Food Restaurant In Sinan

上海思南公馆日式料理

项目名称：上海思南公馆日式料理餐厅, 2010

坐落地点：中国 上海

面积：348m²

设计团队：徐岭啸、汤昊隽、宋原涌、郑琛、孙桢

主要材料：马赛克、古木板、石材

在上海享有盛誉的综合性商业地块——思南公馆，坐落了一座浪漫的日本料理餐厅。这个能容纳60人的餐厅，以强调新鲜的进口原料为主题概念，旨在使用大自然的元素来营造一个休闲的空间。花卉图腾，在空间的墙面和顶面通过各种材料鲜明的诠释着。

餐厅有二楼和三楼两层，两座寿司台及六个包房。三楼另有一个夹层作为办公空间。餐厅的建筑本身是一座总建筑面积不到300平方米精品楼。平面布置合理有效：每一层，入口的右侧是开放式寿司吧台，走道的尽头设有三个包房。每一个包房都可以通过百叶门通往各自布满花朵的小阳台。

Located in the prestigious mixed-use complex Sinan Mansion in Shanghai, this exclusive venue is a romantic Japanese restaurant with a twist. While the main concept of the 60-seat restaurant is to emphasize the freshness of the imported ingredients, the design objective is to create a relaxing space that brings out elements of our natural environment. Floral patterns, through various materials and interpretations, are evident on the walls and ceiling areas of the space.

The restaurant occupies the second and third floors, with two sushi bars and a total of 6 private dining rooms. There is also an additional partial level on the third floor which houses the office area. The restaurant itself is boutique with less than 300 square meters total. The plan layout is efficient and logical: on each floor, the open sushi bar is located right off the entrance and 3 private rooms towards the end of the corridor. Each private dining room has its own shutter doors which open to an intimate balcony lined with flowers.

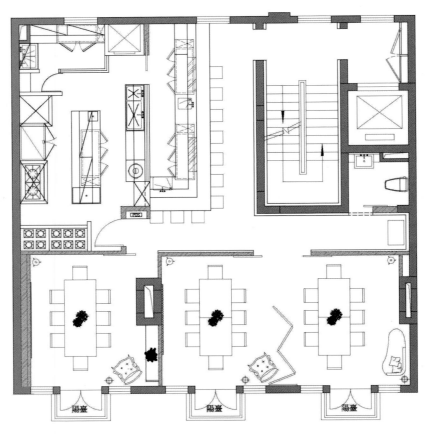

二层平面图/2nd Floor plan

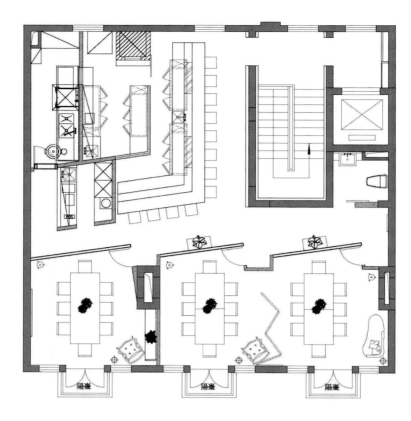

三层平面图/3rd Floor plan

陈德坚

毕业于英国 **De Monfort University** 的室内设计系

德坚设计 创办人

现担任香港室内设计协会 会长

曾任香港设计中心 董事

曾经在曼彻斯特著名的设计公司 **Company Designer Limited** 工作

近期主要荣膺

2009年，获颁 "iF传达设计大奖"

2008年，取得 "Modern Decoration International Media Prize"

2001年，获得 "亚太区室内设计大赛大奖"

The Gold Restaurant

The Gold餐厅

项目名称：*The Gold*

坐落地点：*LKF*，云咸街，香港

面积：**446 m²**

设计：德坚设计（*Kinney Chan & Associates*）

黄金餐厅位于香港中环云咸街，是一个独特且高级的餐厅。餐厅的设计主题是充满活力的彩金。大堂旁边的是一个金色的椭圆形图案的门是一个点缀。性感和标志性的酒吧是用餐空间的亮点之一，它是一个金色巨型的玻璃吊灯，从天花板到一直到地板。另一个特点设计是在用餐区的天花，它是用木头和铜在弯曲的形状下创建的。另一方面露台是由绿化带营造而成。通过设计师的想象力和创造力，黄金餐厅的设计绝对是诱人的和复杂的。

The Gold restaurant is a unique and outstanding fine dining restaurant located in Wyndham Street, Hong Kong. The restaurant takes the colour gold as the vibrant design theme. Next to the dark lift lobby is a golden door embellished with the oval-shaped logo. One of the highlight of the dining space is a sexy and iconic bar; it is like a giant chandelier made of layers of Venetian glass in gold colour down from the ceiling to the floor. Another feature design is a stunning ceiling in the fine dining area; it is created in curved-shape with wood and copper. The terrace on the other hand is hip and fresh, while greenery is being set up, there is also a waterfall and a fireplace in the middle of the pond. Through the imagination and creativity of the designers, the Gold restaurant design appears to be seductive and sophisticated.

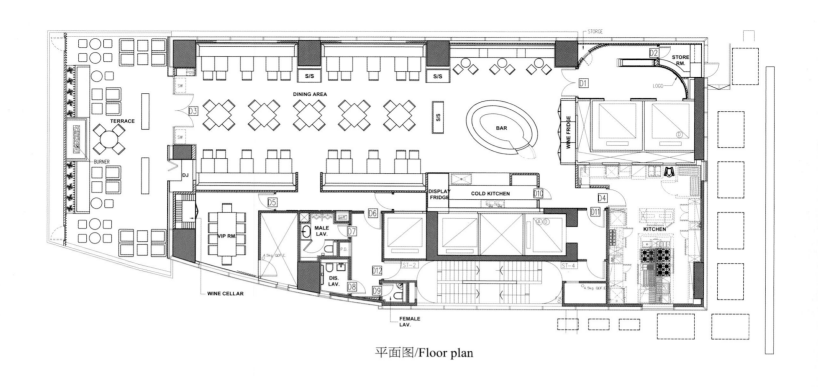

平面图/Floor plan

Haiku Sushi
Haiku寿司

坐落地点：上海国际金融中心

面积：320 m²

设计团队：*Imagine Native Ltd. (Hong Kong)*

主案设计：*Edmond Tse (HKIA), F. Fung (HKIA), C. Lai*

主要材料：石膏板裱糊、橡木复合地板、铝复合板、穿孔铝复合板、透光石、亚麻面料

摄影：*Kingkay*建筑摄影

最新完成的**Haiku**寿司位于上海国际金融中心的一个开放式的庭院中。我们利用折纸的概念，作为整个设计的主要亮点。餐厅空间有如下不同的区域：寿司吧、酒吧，两个主要用餐区，卡座和榻榻米房间。每个区域都形成了一个折纸功，由不同颜色和不同的材料组成，如穿孔铝板复合板，透光石板和亚麻面料。不同的灯光效果，使得每个区域有自己的空间特征。从贵宾室、入口等不同的区域通过天花板上的这些折纸进行区分。它结合了这些功能区，组成了整个餐厅的空间过渡。

The new Haiku Sushi is located at the open courtyard of the recently completed Shanghai International Finance Center. We utilize the concept of origami as the major driven force throughout the design. The restaurant space is sub divided in different zones, sushi bar, drinking bar, two main dining areas, booth seating and tatami rooms. Each zone is formed by an origami feature, which is constructed by colors and different materials, such as perforated aluminum composite panels, translucent stone panels and linen fabrics. With different lighting effects, each zone will have its own spatial character. These different zones in the restaurant are also tied by an origami suspended ceiling from the entrance to the VIP rooms at the end of the shop. It unifies these feature zones and creates a spatial transition throughout the restaurant.

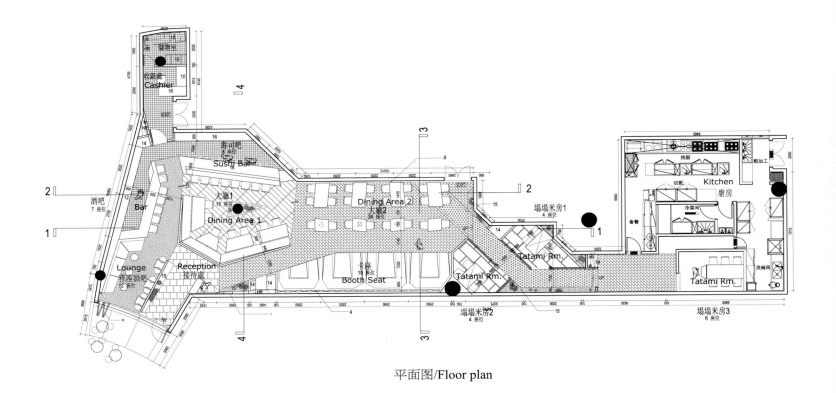

平面图/Floor plan

蔡宗志

中国台北淡江大学 建筑系

1997年，获法国巴黎建筑专业学院建筑师文凭
（Architect DESA ,Paris,France Ecole special
d' architecture ）

2005年，成立法惟思设计工作室（Atelier de
L' A.V.I.S ）

Chasing Dishes and Tappasaki

北京工体享宴火锅铁板烧

坐落地点：北京工体北门

设计公司：北京法惟思设计

设计：蔡宗志

参与设计：王为、詹玉宝、赵志鸿、童孝友

主要材料：木材、方管、石材、水泥

摄影：孙翔宇

餐厅地处北京工体内，餐厅时空背景跨界时空回到北宋北方山水画大师范宽的溪山行旅图中，尝试把中国山水的意境融入空间里，以现代的手法来诠释其树石、树林、婉蜒的流水、人物商旅，屋舍瓦宇等等，再与高大的山体来抗衡，达到空间层次，加上国画中的留白运用，把整体空间分成前中后三段，以中国文人的空间意境来呈现此餐厅的空间层次与文化氛围。

Situated in Beijing Workers' Stadium, the restaurant allows its guests to go back to another space and time with the decorative background painting titled Travelers Amid Mountains and Streams, which was drawn by Fan Kuan, a painter of northern China in Song Dynasty good at painting mountains and waters. The background is intended to involve the artistic conception in Chinese landscape paintings into the restaurant space, interpret the trees, rocks, trees, streams, characters, houses and tiled buildings, which are against the mountains, in the painting by modern means, to achieve a sense of hierarchy, in addition to the space reservation in the painting, which divides the whole space into three section, front, middle and back parts, to present the hierarchization and cultural atmosphere of the restaurant typical of Chinese letters.

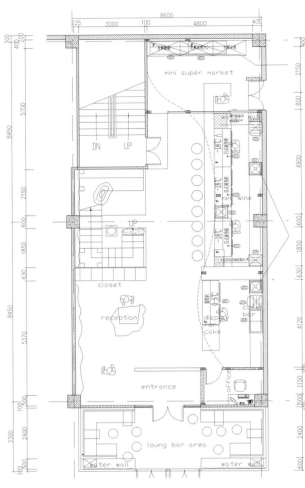

一层平面图/ 1st Floor plan

从入口处至接待处的空间铺陈，让人从繁杂的世俗，慢慢进入一种空灵的感觉，让浮动的心沉淀，紧接着再利用楼高5m的差距，以钢梯插入粗矿大石来点出空间的气势。蜿蜒的流水从上而下经过楼梯扶手而溅在大石上。朔溪而上，成片的树林（包房）与竹林（玻璃酒架）交杂开展，引导进入中段空间，这时人物商旅，热闹非凡，品味之余，无意间视线透过透明厨房而落在对面包房的熟人，举杯共饮。 后段空间，把后厨空间与材料结合作为一种展示融入空间中，如海鲜池里的活海鲜，犹如溪流上的河鲜，挨着散客区，顾客可以从观赏食材到挑选，处理完后而端到面前，作为食材的一种感谢及尊重。进入卫生间的过程中是一种体验。

The layouts from the entrance to the reception allows visitors to overcome hustle and bustle, gradually obtain a spiritual beauty to settle themselves down and then the 5 meters in the height of the building are used by insertion of a steel ladder into a rough rock to create the spacious effect.The serpentine brook flows from top to bottom and splashes itself onto the rock after it passes along the handrail of staircases. Along the brook are trees and bamboos , which are planted here and there and function as the transition to the middle part, which is a busy world filled with the diners.During a feast, it is possible that a diner notice an acquaintance of his/her in the special compartment on the opposite and drink with him/her as the kitchen is transparent. In the back of the restaurant are the kitchen and the area for display of cooking materials,such as live sea food in the pool and river food in the brook, which is adjacent to the tables outside the special compartments, so that the diners are able to view and pick the materials, which shows the gratitude and respects of the restaurant to its customers. Besides, using the bathroom is also a cozy experience.

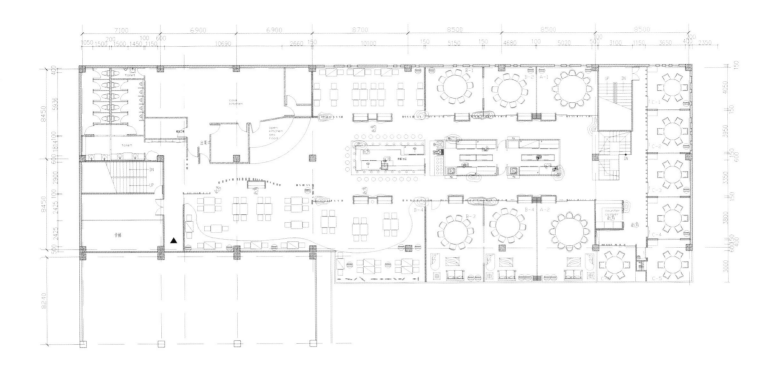

二层平面图/ 2nd Floor plan

利旭恒

英国伦敦艺术大学 **BA**(Hons) 荣誉学士

长年致力于酒店餐饮空间与商业地产的设计工作，10多年的酒店餐饮空间设计，累积了丰富的项目经验，在多个品牌设计风格与商业型态上都卓有成就。

近期主要作品

鼎鼎香

麻辣诱惑

港丽餐厅

烹大师烧肉达人

西堤牛排

王品牛排

豆捞坊

汉拿山

权金城

瑞贝卡

Postscript Restaurant
又及餐厅

坐落地点：北京中关村

项目面积：**850m²**

设计单位：古鲁奇公司

设计师：利旭恒、赵爽、郑雅楠

主要材料：大理石、铝板、地毯、玻璃

摄影：孙翔宇

信写完并已署名后又添上几句，在后添加的几句话下常注明"又及"或"某某又及"。又及餐厅唤起人们对学校餐厅的回忆，柔和的绿色系和天然的大理石提供了一个有机的调色盘，是刚刚踏出校园的学子们填补心灵的加油站，设计师试着提供一个快乐和幸福的空间。

设计师的概念是在600平方米的空间中规划成5个功能区块，除厨房、吧台等基本后场之外，所有的外场用餐区域以环境心理学的模式呈现，每个面向喧嚣都会为主的景观用餐区都被赋予独特的调色。同时，透过窗口静观这纷扰的城市，为不同的人们创造一个属于他们自己的心灵加油站。

After the finish and has signed the letter have added a few words, add a few words in the post under the often marked "p.s.". P.S. restaurant evoke memories of the school cafeteria, soft green color and natural marble system provides an organic palette to people who Just out of school into the community , designer try to provide a space for joy and happiness .

The concept of the designer is planning in the space of 600 square meters into five functional blocks, in addition to the backcourt of the kitchen, bar, etc., all of the outfield dining area showing the pattern of environmental psychology, each facing the hustle and bustlethe landscape will be the main dining area is assigned a unique color. At the same time, through the window to wait and see this turbulent city, for different people to create their own mind gas station.

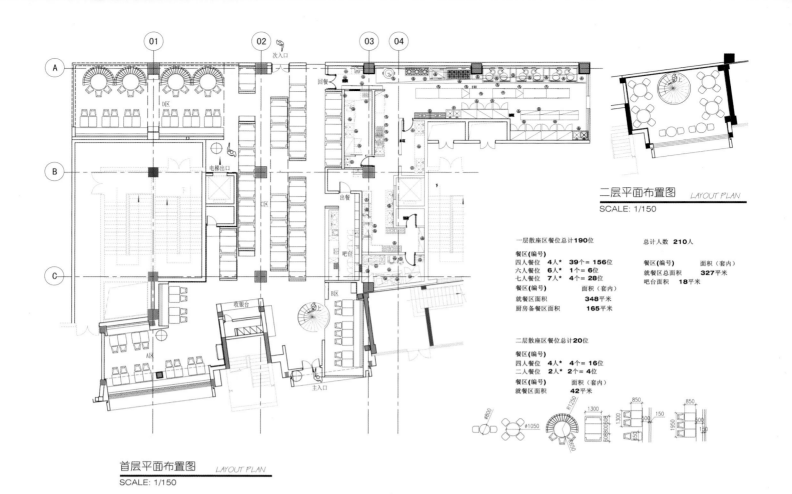

二层平面布置图 *LAYOUT PLAN*

SCALE: 1/150

一层散座区餐位总计190位　　　　　总计人数 210人

餐区(编号)
四人餐位　4人*　39个=156位　　　餐区(编号)　　　面积(套内)
六人餐位　6人*　1个=6位　　　　就餐区总面积　　327平米
七人餐位　7人*　4个=28位　　　　吧台面积　18平米
餐区(编号)　　　面积(套内)
就餐区面积　　348平米
厨房备餐区面积　165平米

二层散座区餐位总计20位

餐区(编号)
四人餐位　4人*　4个=16位
二人餐位　2人*　2个=4位
餐区(编号)　　　面积(套内)
就餐区面积　42平米

首层平面布置图 *LAYOUT PLAN*

SCALE: 1/150

对于现代时尚餐饮空间的设计，顾客心理因素要优先于生理因素来考虑，特别是在繁华的都会中心，用餐当然绝对不只是纯粹的生理行为，更多的是心理学的反射。

设计师针对都会商业区白领族群的用餐心理，精心布局四个属性独特的餐区，各个风格相同手法相异。餐区之间非常注意颜色与材料的运用，小阁楼餐区全绿色空间，白色的楼梯通天隐喻人们努力向上。

Customer psychological factors for the design of the modern and stylish dining space, to give priority to the physiological factors to consider, especially in the bustling city center, dining, of course not just for purely physical act, is a psychological reflection.

Designer for the metropolitan business district, white-collar ethnic dining psychological carefully layout the four unique attributes of the dining area, all style differences in the same practices. Great attention to the use of color and materials between the dining area, small loft dining area of green space, white stairs Babel metaphor that people strive to be up.

悦食I——顶级餐厅设计集成

ISBN 978-7-5038-6092-8
印装：四色精装（320P）
定价：298.00

悦食II——顶级餐厅设计集成

ISBN 978-7-5038-6494-0
印装：四色精装（320P）
定价：320.00

中国室内设计年鉴
——2010厦门特辑

ISBN 978-7-5038-6134-5
印装：四色精装（280P）
定价：260.00

镜像—中国10大样板房设计师

ISBN 978-7-5038-6383-7
印装：四色精装（280P）
定价：256.00

手绘名师表现技法—建筑篇

ISBN 978-7-5038-6264-9
印装：四色平装（168P）
定价：68.00

手绘名师表现技法—室内篇

ISBN 978-7-5038-6263-2
印装：四色平装（168P）
定价：68.00

金设计限量珍藏版(全五册)

·年度优秀酒店、休闲空间
·年度优秀公共·购物空间
·年度优秀样板间/售楼处、办公空间
·年度优秀餐饮、娱乐空间
·年度优秀住宅公寓、别墅

ISBN 978-7-5038-6408-7
开本：230mm×300mm
页码：1500
印装：精装
定价：1288.00元